W9-CCT-585

Introduction to Chemistry for Biology Students

Eighth Edition

George I. Sackheim
Associate Professor Emeritus, Chemistry
University of Illinois at Chicago

formerly
Coordinator of Biological and Physical Sciences
Michael Reese Hospital and Medical Center, Chicago

Science Instructor
St. Francis Hospital School of Nursing
Evanston, Illinois

PEARSON
Benjamin
Cummings

An imprint of Addison Wesley

San Francisco • Boston • New York • Capetown • Hong Kong • London • Madrid
Mexico City • Montreal • Munich • Paris • Singapore • Sydney • Tokyo • Toronto

Publisher: *Jim Smith*
Project Editor: *Trevor Coe*
Managing Producer: *Claire Masson*
Marketing Manager: *Scott Dunstan*
Production Supervisor: *Shannon Tozier*
Production Editor: *Maria McColligan, Nesbitt Graphics, Inc.*
Composition and Illustrators: *Nesbitt Graphics, Inc.*
Manufacturing Buyer: *Robert Davis*
Text Designer: *Patrick Devine*
Cover Illustrator: *Blakeley Kim*
Cover Designer: *Stacy Wong*
Text Printer and Binder: *Phoenix Book Technology Park*
Cover Printer: *Phoenix Book Technology Park*

Credits: Pages 195 and 291: Figures a and b from W. M. Becker and D. W. Deamer, *The World of the Cell*, 3rd ed. © 1996 Benjamin Cummings Publishing Company. Pages 195 and 291: Figures d and e © Irving Geis. Pages 304–307: Appendix figures A, B, C, and D adapted from N. A. Campbell, L. G. Mitchell, and J. B. Reece, *Biology: Concepts and Connections*, 2nd ed. © 1997 Benjamin Cummings Publishing Company.

CIP data is on file at the Library of Congress.

ISBN: 0-8053-3970-1

PEARSON
Benjamin
Cummings

1 2 3 4 5 6 7 8 9 10 – PBT – 07 06 05 04

Contents

To the Student

Introduction to Chemistry for Biology Students, Eighth Edition, is not an ordinary book. It has been programmed to help you review the basic facts, concepts, and terminology of chemistry that are essential to an understanding of biological phenomena. Today's biology courses place increasing emphasis on the chemical processes that underlie the critical biological functions. This book will help you to understand those processes.

The topics involved are among the most critical and exciting that science will explore in the years ahead. What are the basic chemical processes underlying biological phenomena? What are the essential differences between living and nonliving matter? What are the conditions under which molecules organize into living matter? Can these conditions be duplicated experimentally?

If you have already had a course in chemistry, this programmed book can serve as an effective review of the fundamental concepts. If you have had no previous chemistry, the program will give you the background you need to gain a clear understanding of the biological processes you will be studying.

The material covered in this book will help you most if you complete it during the first two weeks of your biology course. Having done that, you will be ready to handle the chemical aspects of biology as they come up.

When you have completed the program, you may want to repeat certain material. To simplify this process, use the book's Index to help you locate specific topics.

If you follow the directions and complete this program, you will learn to

- recognize the elements present in various compounds

- know what is meant by pH and by ionization

- recognize acids, bases, and salts

- discriminate between electrolytes and nonelectrolytes

- understand osmosis and diffusion

- distinguish between passive and active transport

- understand osmotic pressure

- understand transmission of nerve impulses, including depolarization and repolarization

- understand how neurotransmitters work

- know how DNA replicates

- know how mRNA is formed and how it regulates protein synthesis

- understand oxidation and reduction

- know what isotopes are

- recognize various organic functional groups

- distinguish between types of isomers

- differentiate among carbohydrates, fats, and proteins

- understand how enzymes function

- recognize nucleic acids

- understand biologic oxidation, including glycolysis, the Krebs cycle, and the electron transport chain

- understand reactions involved in photosynthesis, including light reactions and the Calvin cycle

- follow the flow of oxygen from the lungs to cells and of carbon dioxide from cells to the lungs on gas partial pressure gradients

How to Use This Book

This type of instructional book may be new to you. Its subject matter has been presented as a series of numbered problems. Each builds on information learned in the preceding problems. For that reason, it is important not to skip around. The sequence of the problems is important because it is programmed to help you learn more efficiently.

Respond to Every Problem

Some problems present new information; others review material presented earlier. Every problem presents a learning task that requires some response from you.

You may be asked to make any of the following types of responses:

❦ write an answer in a blank space

❦ label a diagram

❦ draw a simple diagram

❦ select the correct answer from among several alternatives

❦ write a sentence in answer to a question

When you have written or marked your answer, you will want to find out whether you are correct. Programmed instruction provides you with important feedback by giving you easy access to the answers, which are located in shaded areas on the outer side of each page. *Do not look at the correct answer until after you have marked your own answer.* If you look before answering, you will only impair your own learning process.

Use an Answer Mask

Bound into the back of the book is a sheet of heavy paper that is perforated. Tear off the outer portion of the sheet for use as an answer mask. Here is what to do:

1. As you start working on the first page, a right-hand page, cover the shaded answer column with the answer mask *before* looking at the problems. When you turn the page, shift your answer mask to cover the answer column on the left-hand page before reading the problems on that page. *Be sure you have covered the answers before you read anything.*

2. Each problem number is centered on the page with a shaded rectangle behind it. Read the problem carefully, then record your answer. Make sure you either write each answer or do whatever the directions say. Do not simply think of the answer and then go on. Actually writing or marking your answer reinforces your learning.

3. Now move the answer mask aside to reveal the answer, which you will find aligned with the problem number. (If the main part of the problem runs onto the next page, you will find the answer at the top of that page.)

4. If your answer was correct, move on to the next problem.

5. If your answer was incorrect, reread the problem (if necessary, reread several of the preceding problems) until you understand your error and know why the given answer is correct. Then go on.

When you have completely worked your way through this book, you should have the knowledge of chemistry you need to succeed in your biology courses.

To find a topic or to review a topic already completed, look in the index to locate the reference(s) to that subject.

PART **I**

Inorganic Chemistry

Atomic Structure 1

ELEMENTARY PARTICLES

Atoms are made up of several components. Collectively these components are called the *elementary particles*. We will be discussing the three major elementary particles: *protons, neutrons*, and *electrons*.

Here is a diagram of an atom:

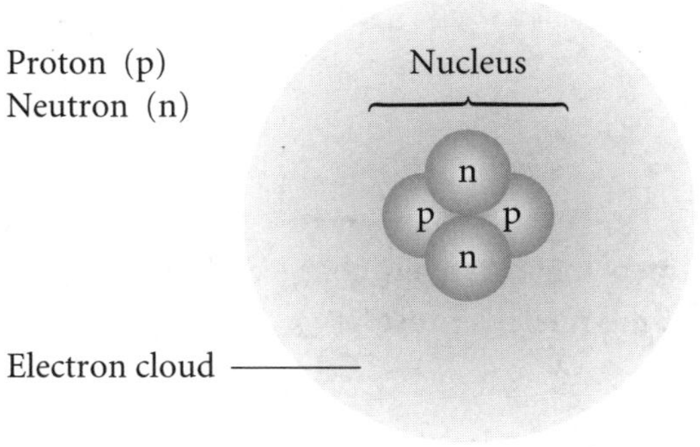

Proton (p)
Neutron (n)

Nucleus

n
p p
n

Electron cloud ————

1

The protons (p) and the neutrons (n) are packed together in an inner core called the _____. The outer part of the atom, which contains electrons, is called the _____.

a negative electrical
charge, because the
electron cloud consists of
electrons

2

The electron cloud has a negative electrical charge. What type of charge would you expect the electron to have? _____

_____

a. repel
b. attract
c. repel
d. attract

3

The electron has a *negative electrical charge* and is symbolized by e^-. Remember that *like* electrical charges repel each other and *unlike* charges attract.

Indicate whether the following pairs of charges would attract or repel each other.

a. ⊕ ⊕ _____   b. ⊖ ⊕ _____

c. ⊖ ⊖ _____   d. ⊕ ⊖ _____

positive

4

The nucleus attracts the negatively charged electrons. Therefore, the overall charge of the nucleus must be _____ (negative/positive).

proton

5

The neutron was named for its electrical characteristics. It has no electrical charge; it is neutral. In other words, the positive charge of the nucleus must be due to the *second type of particle* it contains. This second type of particle is the _____.

6

So far, then, we have this picture of atomic structure:

a.   An atom consists of an inner part, or _____, that is made up of _____ and _____.

b.   The electron has what type of charge? _____

c.   The proton has what type of charge? _____

d.   The neutron has a charge of _____.

7

The charge on the electron balances the charge on the proton. If the electron has a charge of −1, then the proton would have a charge of _____ $(-1, +1, \pm 1)$.

8

An atom with one proton in its nucleus and one electron outside that nucleus would therefore have an overall charge of _____ $(+1, -1, 0)$.

9

Atoms are electrically neutral. Therefore, an atom will contain: (check one)

___ more protons than electrons
___ more electrons than protons
___ an equal number of protons and electrons

a. nucleus; protons; neutrons
b. negative
c. positive
d. zero (0)

+1

0

an equal number of protons and electrons

12

a. proton
b. electron

2; 2; 2

10

An atom with 12 protons in the nucleus would have how many electrons outside the nucleus? _____

11

The atom with the simplest atomic structure is hydrogen. For simplicity we shall merely indicate the electron(s) outside the nucleus and omit the electron cloud.

Hydrogen

a.   The nucleus of the hydrogen atom consists of one _____.

b.   The outer part of the atom, the electron cloud, contains one

_____.

12

The helium atom is a little more complicated.

$$e^- \quad \left(\begin{array}{c} n \\ p \quad p \\ n \end{array} \right) \quad e^-$$

Helium

It contains: (how many?)

_____ neutrons
_____ protons
_____ electrons

ATOMIC NUMBER

There are over 100 known elements. Each element has two numbers associated with it, numbers that give certain facts about the structure of its atoms.

The first number is the *atomic number*. The atomic number is the number of protons in the nucleus of the atom.

13

Hydrogen, the simplest atom, contains only one proton, so the atomic number of hydrogen is _____.

1

14

Uranium is the most complicated of the elements that occur naturally. A uranium atom contains 92 protons, 146 neutrons, and 92 electrons. The atomic number of uranium is _____.

92

15

An atom of magnesium, atomic number 12, must have a nucleus containing _____ protons.

If the nucleus contains 12 protons, then there must be how many electrons? _____

12; 12

16

Therefore, the atomic number of an element indicates the number of _____ in the nucleus of the atom and also the number of _____ outside the nucleus.

protons; electrons

MASS NUMBER *

The second number associated with each atom is the *mass number*. The mass number expresses the sum of the masses of the particles in the atom.

A proton has a mass of 1 dalton. An electron is considered to have zero mass, or a mass of 0.

1

17

A hydrogen atom has a mass of _____. (If you don't know, see problem 11.)

a. 2
b. 0
c. 2
d. 1

18

The helium atom has a mass number of 4.

a.   The 2 protons in the helium atom have a total of how many daltons? _____

b.   The 2 electrons in the helium atom have a total of how many daltons? _____

c.   Therefore, for the helium atom to have a mass number of 4, the 2 neutrons must contain how many daltons? _____

d.   If 2 neutrons have a total of 2 daltons, a neutron must have an atomic mass of _____.

* The *atomic mass* of an atom is the weighted average of the masses of its isotopes (see the later section on isotopes). The atomic mass of an atom is a decimal value, as shown in the periodic table on the inside front cover of this book. The *mass number* of an atom is a whole number and is equal to the sum of protons and neutrons in that atom.

19

Because the electrons, which have practically no mass, are located outside the nucleus, the entire mass of the atom can be considered to be located:

___ in its electron cloud   ___ in the nucleus

20

The *atomic number* indicates the number of protons (each with atomic mass 1) inside the nucleus of an atom. The *mass number* indicates the number of protons and neutrons (each with atomic mass 1) in the nucleus. Therefore, the number of neutrons can be determined by *subtracting* the atomic number from the mass number.

The sodium atom has a mass number of 23 and an atomic number of 11. The number of neutrons in the nucleus of the sodium atom is

_____.

21

The carbon atom has an atomic number of 6 and a mass number of 12. The carbon atom contains: (how many?)

_____ protons in its nucleus
_____ neutrons in its nucleus
_____ electrons outside its nucleus

22

The element phosphorus has an atomic number of 15 and a mass number of 31. Indicate on the blank lines in the diagram the number of protons, neutrons, and electrons.

Phosphorus

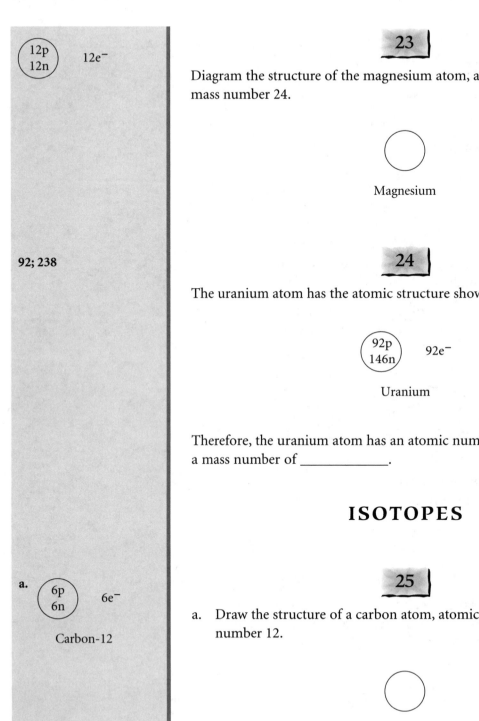

12p
12n   12e⁻

92; 238

23

Diagram the structure of the magnesium atom, atomic number 12 and mass number 24.

◯

Magnesium

24

The uranium atom has the atomic structure shown here.

92p
146n   92e⁻

Uranium

Therefore, the uranium atom has an atomic number of _____ and a mass number of _____.

ISOTOPES

a.

6p
6n   6e⁻

Carbon-12

25

a. Draw the structure of a carbon atom, atomic number 6 and mass number 12.

◯

b. Draw the structure of a carbon atom, atomic number 6 and mass number 13.

26

Here are the structures you drew for the two carbon atoms:

Carbon-12   Carbon-13

a. These atoms have _____ (the same/different) atomic number(s).

b. These atoms have _____ (the same/different) mass number(s).

Such atoms are called *isotopes*.

27

Isotopes, then, may be defined as atoms that have:

___ the same atomic number and the same mass number
___ different atomic numbers
___ different mass numbers and the same atomic number

b.

6p
7n   6e⁻

Carbon-13

a. the same
b. different

different mass numbers
and the same atomic
number

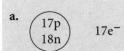

Chlorine-35

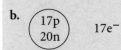

Chlorine-37

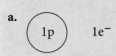

Hydrogen-1

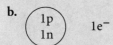

Hydrogen-2

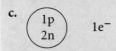

Hydrogen-3

a. **same**
b. **different**

28

Draw the two isotopes of chlorine, atomic number 17 and mass numbers 35 and 37.

a.   chlorine-35   b.   chlorine-37

29

Draw the three isotopes of hydrogen, atomic number 1 and mass numbers 1, 2, and 3.

a.   hydrogen-1   b.   hydrogen-2   c.   hydrogen-3

30

Radioisotopes, isotopes that give off radiation, are frequently used in medical applications. I-131 (radioactive iodine, also written as ^{131}I) is used in the diagnosis and treatment of thyroid conditions. How does I-131, radioactive iodine, compare with I-127, nonradioactive iodine, in

a.   atomic number _____

b.   mass number _____

ELECTRON ENERGY LEVELS

The electrons are located outside of the nucleus of the atom. These electrons make up the electron cloud, which may be subdivided into different energy levels. The first energy level is nearest the nucleus; then comes the second energy level, the third energy level, and so on.

Each energy level can hold a certain maximum number of electrons. This maximum number may be determined by using the formula $X = 2n^2$ (*X* is the maximum number of electrons in energy level number *n*).

31

2; 2

Using the formula $X = 2n^2$, if $n = 1$, then $X =$ _____. The energy level indicated by $n = 1$ is the first energy level. Therefore, the first energy level can hold a maximum of _____ electrons.

32

a. yes
b. no

a. Can the first energy level hold fewer than two electrons? _____

b. Can the first energy level hold more than two electrons? _____

33

8

For the second energy level, where $n = 2$, the maximum number of electrons is _____.

34

a. 18
b. 32

a. The maximum number of electrons in the third energy level is

_____.

b. The maximum number of electrons in the fourth energy level is

_____.

2e⁻ in the first energy
level; 8e⁻ in the second
energy level; 18e⁻ in the
third energy level; 32e⁻
in the fourth energy level

8

35

Label the maximum number of electrons possible in each energy level in
the diagram.

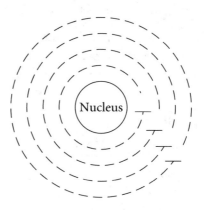

36

The following rules must be observed when considering the placement
of electrons in the various energy levels. The first energy level must be
filled with 2 electrons before electrons can go into the second energy
level.

The second energy level must be filled with _____ electrons
before electrons can go into the third energy level.

For elements having more than three energy levels, the sequence of
filling those levels is complex and can be found in a general chemistry
text.

Hydrogen atom

a. 1
b. 0
c. 1
d. **first**

37

On the diagram, show the structure of the hydrogen atom, atomic number 1 and mass number 1.

Hydrogen atom

a. The number of protons in the hydrogen atom is _____ .

b. The number of neutrons is _____ .

c. The number of electrons is _____ .

d. The 1 electron in the hydrogen atom must go into which energy level: first, second, or third? _____

38

In the space provided, draw the structure of the helium atom, atomic number 2 and mass number 4.

Helium atom

Helium atom

a. 3
b. 4
c. 3

2

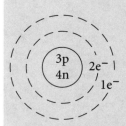

Lithium atom

39

Now let's look at the structure for the lithium atom, atomic number 3 and mass number 7.

a.   The number of protons in the lithium atom is _____.

b.   The number of neutrons in the lithium atom is _____.

c.   The number of electrons in the lithium atom is _____.

40

There are 3 electrons in the lithium atom. How many energy levels will the lithium atom have? _____ (If you aren't sure, check problem 36.)

41

Complete the diagram of the lithium atom.

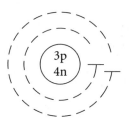

Lithium atom

42

The carbon atom has atomic number 6 and mass number 12.

a. The carbon atom contains _____ protons.

b. The carbon atom contains _____ neutrons.

c. The carbon atom contains _____ electrons.

d. Complete the structure of the carbon atom.

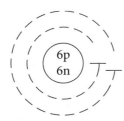

Carbon atom

43

a. In the element sodium, atomic number 11, there are how many electrons? _____

b. How many energy levels will the sodium atom have? _____

c. How many electrons will each energy level of the sodium atom have?

1st energy level __   2nd energy level __   3rd energy level __

d. Diagram the complete structure of the sodium atom, atomic number 11 and mass number 23.

Sodium atom

a. 6
b. 6
c. 6
d.

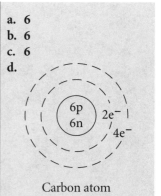

Carbon atom

a. 11
b. 3
c. 2; 8; 1

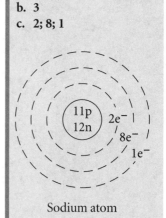

Sodium atom

Silicon atom

Diagram the structure of the silicon atom, atomic number 14 and mass number 28.

Chemical Symbols 2

Elements are the building blocks of the universe. They cannot be broken down into simpler substances by ordinary chemical means. Each of the over 100 elements has its own name and its own properties. The symbol for an element is usually an abbreviation for its name. Frequently the abbreviation is merely the first letter of that name. The following table lists some of these elements.

Element	Symbol
Hydrogen	H
Carbon	C
Oxygen	O
Nitrogen	N
Phosphorus	P
Sulfur	S
Iodine	I

45

What elements are present in water, H_2O? _____

hydrogen and oxygen

46

What elements are present in ammonia, NH_3? _____

nitrogen and hydrogen

carbon, hydrogen, and oxygen

47

What elements are present in glucose, $C_6H_{12}O_6$? _____

carbon, hydrogen, oxygen, nitrogen, sulfur

48

Biotin, one of the B-vitamins, $C_{10}H_{16}O_3N_2S$, contains which elements? _____

carbon, hydrogen, oxygen, nitrogen, phosphorus

49

Adenosine triphosphate, ATP, the body's primary energy compound, has the formula $C_{11}H_{18}O_{13}N_5P_3$. What elements does it contain? _____

Ca; Br; Si; Ba

50

When the names of more than one element begin with the same letter, frequently the second letter of the name is added to the symbol. Note that only the first letter of the symbol is capitalized.

Here are some of the elements whose symbols are the first two letters of their names. Write the symbols for each.

Element	Symbol
Calcium	_____
Bromine	_____
Silicon	_____
Barium	_____

calcium and bromine

51

$CaBr_2$ contains which two elements? _____

52

BaO$_2$ contains which two elements? _____

barium and oxygen

53

CaC$_2$O$_4$ contains which three elements? _____

calcium, carbon, and oxygen

There are several elements whose symbols are not derived from the first letter or first two letters of their English names. Some of those elements are listed in the following table.

Element	Symbol
Chlorine	Cl
Sodium	Na (from the Latin *natrium*)
Magnesium	Mg
Potassium	K (from the Latin *kalium*)
Zinc	Zn
Iron	Fe (from the Latin *ferrum*)

54

Write the name of the element represented by each of the following symbols:

H _____   Na _____   Si _____
N _____   K _____   Br _____
O _____   C _____   Zn _____
Ca _____   S _____   Fe _____

Hydrogen; Nitrogen; Oxygen; Calcium; Sodium; Potassium; Carbon; Sulfur; Silicon; Bromine; Zinc; Iron

a. potassium and
   bromine
b. carbon and chlorine
c. sodium, oxygen, and
   hydrogen
d. magnesium and
   chlorine
e. sodium, nitrogen,
   and oxygen
f. potassium, iron,
   carbon, and nitrogen

55

What elements are present in each of the following?

a.　KBr _____

b.　CCl_4 _____

c.　NaOH _____

d.　$MgCl_2$ _____

e.　$NaNO_3$ _____

f.　$K_3FeC_6N_6$ _____

a. magnesium and
   nitrogen

56

In addition to carbon, hydrogen, and oxygen, which elements are present in the following compounds whose structures are indicated?

a.　chlorophyll a _____

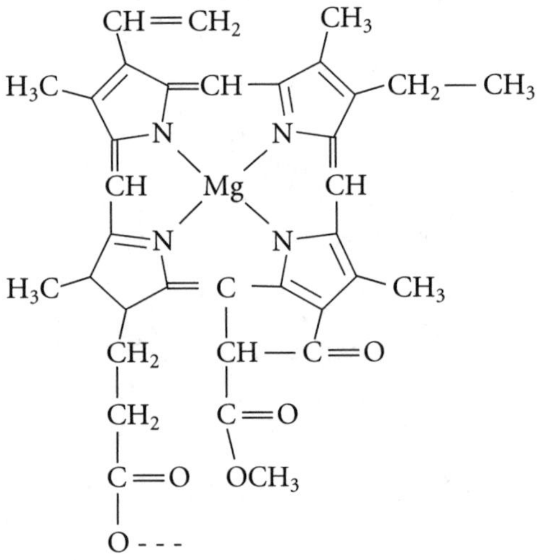

b.   heme (a part of the hemoglobin molecule) _____

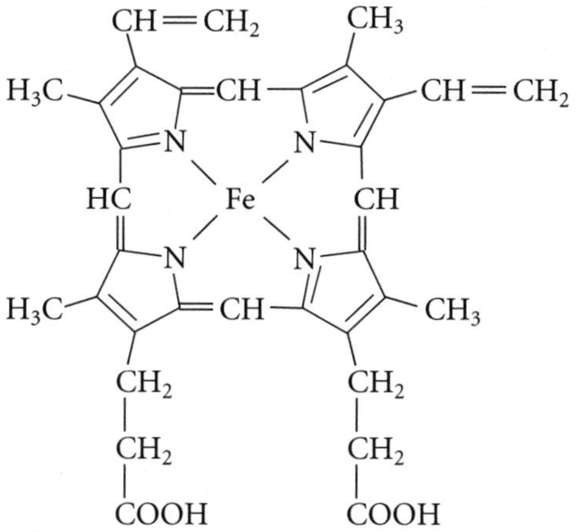

b. iron and nitrogen

c.   T$_4$, a thyroid hormone _____

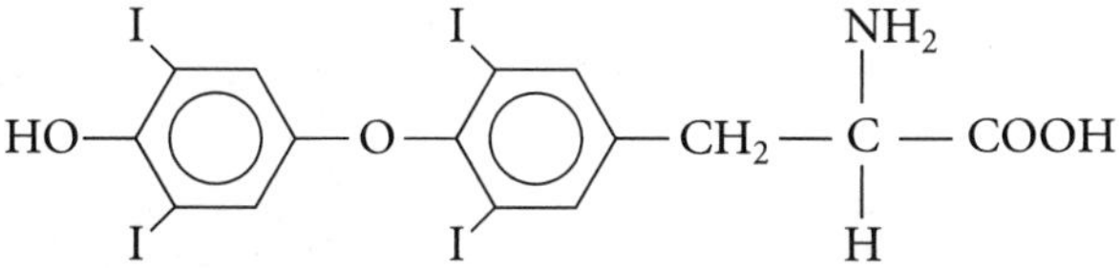

c. iodine and nitrogen

d.   cysteine, a non-essential aminoacid _____

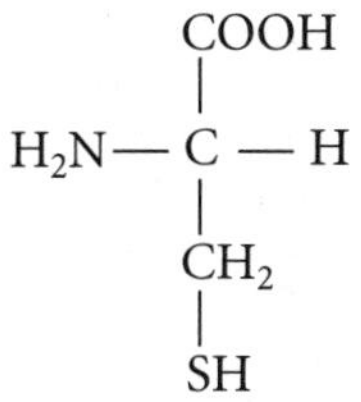

d. sulfur and nitrogen

Atoms and Molecules 3

57

One atom of hydrogen is written as H. The number 1 is understood. Two atoms of hydrogen are written as 2H. Three atoms of carbon are written as _____.

3C

58

Atoms combine to form *molecules*. A molecule contains two or more atoms of the same or different elements. The molecule O_2 contains how many atoms of oxygen? _____

2

59

Each molecule of glucose, $C_6H_{12}O_6$, contains:

_____ atoms of carbon
_____ atoms of hydrogen
_____ atoms of oxygen

6; 12; 6

60

A water molecule is composed of two hydrogen atoms and one oxygen atom. Write the formula for a water molecule. _____

H_2O

$C_{12}H_{22}O_{11}$

$C_{63}H_{90}O_{14}N_{14}PCo$

a. 6
b. $8H_2O$

molecule; hydrogen; 2

61

A molecule of sucrose contains 12 atoms of carbon, 22 atoms of hydrogen, and 11 atoms of oxygen. The formula for sucrose is _____.

62

One molecule of vitamin B_{12} contains:

63 atoms of carbon
90 atoms of hydrogen
14 atoms of oxygen
14 atoms of nitrogen
 1 atom of phosphorus
 1 atom of cobalt (Co)

The formula for vitamin B_{12} is _____.

63

When you want to represent more than one molecule of a compound, the number of molecules precedes the symbols.

a.   Thus, $6CO_2$ indicates how many molecules of carbon dioxide?

_____

b.   How would eight molecules of water be represented? _____

64

The formula H_2 indicates one _____ of _____, composed of how many atoms? _____

65 |

The formula $2H_2$ indicates _____ of hydrogen.

two molecules

66 |

What do the following formulas represent?

3O _____ $4N_2$ _____ $7H_2O$ _____

3 atoms of oxygen;
4 molecules of nitrogen;
7 molecules of water

Ionization 4

IONS

67

Atoms contain equal numbers of protons and electrons. Thus, atoms are:

__ electropositive   __ electronegative   __ electrically neutral

68

The electron is a _____ (negatively/positively) charged particle. So, if an atom loses an electron, it will then have an overall _____ (negative/positive) charge.

69

If an atom were to *gain* an electron, it would then have an overall _____ charge.

70

An *ion* is an atom that has acquired an electrical charge by either losing or gaining a(n) _____.

positively

Note that electrons, which are located outside the nucleus, are lost or gained in the formation of ions. The nucleus, which contains protons and neutrons, does not take part in the formation of ions.

If a hydrogen atom loses its electron, it becomes a hydrogen ion. Because it loses an electron, it is a _____ (negatively/positively) charged ion.

a. ion
b. negatively
c. positively

a.   Any atom that has lost or gained an electron is an _____.

b.   If an atom gains an electron, it becomes a _____ charged ion.

c.   If an atom loses an electron, it becomes a _____ charged ion.

valence shell

Some elements reach their most stable state when they have 2 electrons in their outer (their only) energy level, which is called the *valence shell*.

Helium atom

This atom of helium, for example, contains 2 electrons, which fill the first energy level. So, the first energy level in helium is called the

_____.

stable

74

Helium (He) has a completed valence shell, one that can hold no more electrons. Thus, helium is a very _____ (stable/unstable) atom. Such elements are usually unreactive.

valence shell

75

Most atoms reach their most stable state when they have 8 electrons in their outer energy level. This property is called the octet rule.

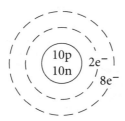

Neon atom

This atom of neon (Ne), for example, contains 10 electrons. Two of the electrons fill the first energy level, and the remaining 8 electrons are in the second energy level. So the second energy level of neon would be called its _____.

a. valence shell; third; 1
b. Cl

76

Look at this atom of chlorine.

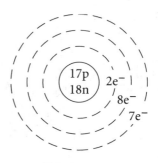

Chlorine atom

a. In chlorine, the outer energy level, or _____, is the
    _____ energy level. It contains 7 electrons. Thus, to attain
    a stable state, the chlorine atom needs to gain how many electrons?
    _____

b. What is the symbol for chlorine? _____

a. ion
b. negative

77

Suppose that an atom of chlorine gains the 1 electron it needs to become
stable.

a. Because it has gained an electron, it is now a charged atom, called
    a(n) _____.

b. Would it have a positive or negative charge? _____

Cl^-

78

Which of the following symbols stands for the chloride ion (a chlorine
atom that has gained 1 electron)?

___ Cl   ___ Cl^+ ___ Cl^-

79

If a chlorine atom is to gain an electron, some other atom must lose the electron.

a. That atom could be sodium, the symbol for which is _____.

b. The sodium atom has how many electrons in its valence shell?

_____

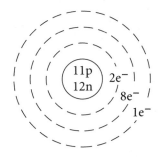

Sodium atom

c. To become stable it will have to _____ (gain/lose) one electron.

80

When the sodium atom loses 1 electron, it becomes a _____ (positively/negatively) charged _____.

81

Which symbol should we use for the sodium ion?

___ Na   ___ Na$^+$ ___ Na$^-$

a. Na
b. 1
c. lose

positively; ion

Na$^+$

a. +1
b. −1
c. +2

82

Positive and negative charges are indicated by superscript plus (positive) or minus (negative) signs, respectively, with the number 1 being understood. Charges greater than 1 are written as superscripts, with the number preceding the plus or minus sign.

a. Na^+ indicates a sodium ion with a charge of _____.

b. Cl^- indicates a chloride ion with a charge of _____.

c. Mg^{2+} indicates a magnesium ion with a charge of _____.

a. Al^{3+}
b. S^{2-}

83

a. How would an aluminum ion with a charge of $+3$ be indicated?

    _____

b. How would a sulfide ion with a charge of -2 be indicated?

    _____

loses; positive; 1;
gains; negative; 1

84

$$Na + Cl \longrightarrow Na^+ + Cl^-$$

Reaction of sodium and chlorine

When a sodium atom reacts with a chlorine atom, the sodium atom _____ (gains/loses) 1 electron to form a sodium ion with a charge of _____ (how many?).

At the same time, the chlorine atom _____ (gains/loses) 1 electron to form an ion with a _____ charge of _____ (how many?).

85

Look at this magnesium atom. Its symbol is _____.

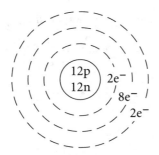

Magnesium atom

It has _____ electrons in its valence shell. So, to gain
stability, it will have to _____ (gain/lose) _____
(how many?) electrons.

The symbol for the magnesium ion is _____ (Mg^{2-}, Mg,
Mg^{2+}).

Mg; 2; lose; 2; Mg^{2+}

86

The reaction between magnesium and chlorine may be written as:

$$Mg + Cl_2 \longrightarrow Mg^{2+} + 2Cl^-$$

This reaction shows that 1 chlorine molecule reacts with 1 magnesium
atom to form 2 chloride ions, each with a charge of _____, and
1 magnesium ion with a charge of _____.

$-1; +2$

a. 3
b. 6
c. −2
d. S^{2-}

87

a. The sulfur atom, atomic number 16, has how many energy levels? _____

b. How many electrons are in its valence shell? _____

c. If a sulfur atom gains 2 electrons to fill its valence shell, then it will have a charge of _____.

d. The symbol for the sulfide ion is _____.

88

How can we tell whether an atom will lose or gain electrons to reach a stable structure of 8 electrons in its valence shell? Look over the following diagrams.

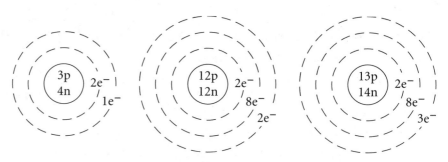

Form positive ions

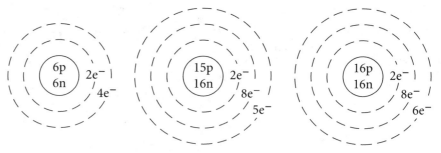

Do not form positive ions

In general, an atom with _____ (how many?) electrons in
its valence shell tends to lose electrons and form a _____
(positively/negatively) charged ion.

89

Sodium has atomic number 11. How many electrons does it have in its
valence shell? _____

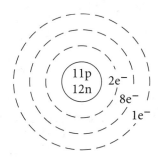

Sodium atom

Therefore, the sodium ion is formed by the _____ (loss/gain)
of 1 electron.

The symbol for the sodium ion is _____.

90

Here is the hydrogen atom, atomic number 1 and mass number 1.

Hydrogen atom

Hydrogen usually loses its 1 electron to form a hydrogen ion with a
charge of _____.

The hydrogen ion can be represented as _____.

3; loss; Al^{3+}

91

Here is an aluminum atom:

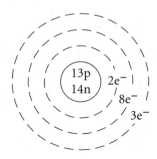

Aluminum atom

Its valence shell contains _____ electrons. The aluminum ion is formed by the _____ (loss/gain) of 3 electrons.

The symbol for the aluminum ion is _____.

Mg^{2+}

92

Magnesium, atomic number 12, forms a magnesium ion that may be represented as _____.

93

Look at the following diagrams carefully.

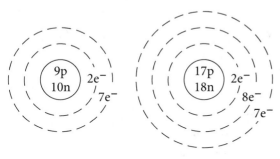

Form negative ions

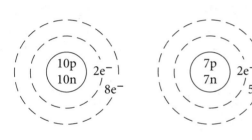

Do not form negative ions

In general, atoms that have _____ (how many?) electrons in
their valence shells form negatively charged ions.

94

When an atom gains electrons to fill its valence shell, it forms _____
charged ions.

negatively

9; 7; gain; 1; F⁻

Complete the following for the fluorine (F) atom shown here:

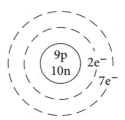

Fluorine atom

Atomic number: _____
Number of electrons in valence shell: _____
Ion formed by _____ (gain/loss) of _____ electron(s)
Symbol for ion: _____

S; 16; 32; S²⁻

Complete the following for the sulfur atom shown here:

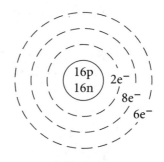

Sulfur atom

Symbol for sulfur: _____
Atomic number: _____
Mass number: _____
Symbol for ion: _____

97

Atoms with 4 or 5 electrons in their valence shells usually do not form ions. We shall discuss these ions later.

Atoms with 8 electrons in their valence shells are stable. Because they do not have to gain or lose electrons, you would expect these stable atoms to be:

__ reactive   __ inert (unreactive)

inert (unreactive)

98

The sodium atom, atomic number 11, has how many electrons in its valence shell? _____

The sodium atom will tend to lose the 1 electron in its valence shell to form an ion with a charge of _____.

1; +1

99

The chlorine atom, which has the symbol _____, has atomic number 17. It has how many electrons in its valence shell? _____

The chlorine atom will tend to gain _____ (how many?) electron(s) to form an ion with a charge of _____.

Cl; 7; 1; −1

IONIC BONDS

gains

100

Consider the reaction between sodium and chlorine:

$$Na + Cl \longrightarrow Na^+ + Cl^-$$

The sodium ion and the chloride ion are oppositely charged. These ions are held together by the attraction of their opposite charges. We say that there is an *ionic bond* between the sodium ion and the chloride ion.

An ionic bond is produced whenever one atom loses an electron or electrons and another atom _____ an electron or electrons.

ionic bond

101

In the reaction $Zn + S \longrightarrow Zn^{2+} + S^{2-}$, the zinc ion and the sulfur ion are held together by a(n) _____.

ion; negatively; ion

102

When a sodium atom combines with a chlorine atom according to the equation $Na + Cl \longrightarrow Na^+ + Cl^-$, a compound containing a positively charged sodium _____ and a _____ charged chloride _____ is formed. This compound (sodium chloride) is usually written as NaCl, with the charges being understood and not written.

chloride; Cl^-

103

When the compound NaCl is placed in water, the ionic bond holding the sodium ion and the chloride ion together is weakened, so that these ions are free to move throughout the solution.

Therefore, NaCl in water produces a *solution* containing sodium ions (Na^+) and _____ ions (_____).

ACIDS, BASES, AND SALTS

104

a. When the compound HCl is placed in water, it produces a solution containing _____ ions (_____) and _____ ions (_____).

b. When the compound H_2SO_4 is placed in water, it produces a solution containing _____ ions and HSO_4^- ions.

a. **hydrogen; H^+; chloride; Cl^-**
b. **hydrogen (H^+)**

105

Any substance that yields hydrogen ions (H^+) in solution is called an *acid*.

$$HCl \longrightarrow H^+ + Cl^-$$
$$H_2SO_4 \longrightarrow H^+ + HSO_4^-$$
$$HNO_3 \longrightarrow H^+ + NO_3^-$$

HNO_3 is an *acid* because it yields _____ in solution.

hydrogen ions or H^+

106

A hydrogen ion is also a *proton*. Therefore, acids yield hydrogen ions or _____ in solution.

protons

acids

107

Note that hydrogen ions associate with water molecules to form a "hydronium" ion $(H_3O)^+$. Nonetheless, we will use the term *hydrogen ion* (H^+) in this book.

Substances that yield hydrogen ions or protons in solution are called _____. Substances that accept (react with) hydrogen ions or protons are called *bases*.

bases

108

When sodium hydroxide, NaOH, is placed in water, sodium ions and hydroxide ions are present:

$$NaOH \longrightarrow Na^+ + OH^-$$

The hydroxide ions (OH^-) react with hydrogen ions to form water:

$$OH^- + H^+ \longrightarrow H_2O$$

Because hydroxide ions accept (react with) hydrogen ions, OH^- ions are _____.

yes

109

Potassium hydroxide, KOH, also yields OH^- ions in water. Would KOH be a base? _____

bases; they accept (react with) hydrogen ions

110

Bicarbonate ions, HCO_3^-, are important ions in body fluids. They react as follows:

$$HCO_3^- + H^+ \longrightarrow H_2CO_3$$

Are bicarbonate ions acids or bases? _____

Why? _____

111

Ammonia is a waste product of the body's metabolism of protein. Ammonia reacts as follows:

$$NH_3 + H^+ \longrightarrow NH_4^+$$

Ammonia   Ammonium ion

Is ammonia an acid or a base? _____

base

112

Acids _____ hydrogen ions or protons.

Bases _____ hydrogen ions or protons.

yield; accept (react with)

113

When the compound NaCl is placed in water, it yields _____ ions and _____ ions.

sodium (or Na$^+$); chloride (or Cl$^-$)

114

The solution of NaCl in water would not be considered an acid because:

___ it yields OH$^-$ ions
___ it doesn't break up into ions
___ it yields no hydrogen ions

it yields no hydrogen ions

a. salt
b. acid
c. salt
d. acid
e. base
f. acid
g. base

115

A compound that yields ions other than hydrogen ions (H^+) or hydroxide ions (OH^-) is called a *salt*. Are the following underlined substances acids, bases, or salts?

a. $\underline{Na_2CO_3} \longrightarrow 2Na^+ + CO_3^{2-}$ _____

b. $\underline{H_2SO_4} \longrightarrow H^+ + HSO_4^-$ _____

c. $\underline{MgCl_2} \longrightarrow Mg^{2+} + 2Cl^-$ _____

d. $\underline{H_2PO_4^-} \longrightarrow H^+ + HPO_4^{2-}$ _____

e. $\underline{NaOH} + H^+ \longrightarrow Na^+ + H_2O$ _____

f. $\underline{H_3PO_4} \longrightarrow H^+ + H_2PO_4^-$ _____

g. $\underline{Ca(OH)_2} + 2H^+ \longrightarrow Ca^{2+} + 2H_2O$ _____

ELECTROLYTES
AND NONELECTROLYTES

yes

116

Acids, bases, and salts are called *electrolytes*. Solutions of electrolytes conduct electricity because of the presence of ions:

$$HCl \longrightarrow H^+ + Cl^-$$
Hydrochloric
acid

Would a solution of hydrochloric acid be an electrolyte? _____

117

Here's what happens when magnesium chloride is put in solution:

$$MgCl_2 \longrightarrow Mg^{2+} + 2Cl^-$$

a. Does the solution of $MgCl_2$ contain ions? _____

b. Is $MgCl_2$ an acid, a base, or a salt? _____

c. Is a solution of $MgCl_2$ an electrolyte? _____

118

When sucrose, $C_{12}H_{22}O_{11}$, is placed in water, no ions are produced.

a. Would a solution of sucrose conduct electricity? _____

b. Would a solution of sucrose be an electrolyte or a nonelectrolyte?

_____

119

Which of the following solutions would be electrolytes?_____

a. $MgSO_4 \longrightarrow Mg^{2+} + SO_4^{2-}$

b. alcohol $\longrightarrow$ no ions

c. $KOH \longrightarrow K^+ + OH^-$

d. $C_6H_{12}O_6$ (glucose) $\longrightarrow$ no ions

e. $HNO_3 \longrightarrow H^+ + NO_3^-$

a. yes
b. salt
c. yes

a. no
b. nonelectrolyte

a, c, e

ANIONS AND CATIONS

120

Consider the following electrolytic system, a battery whose two electrodes are immersed in a solution containing the indicated ions.

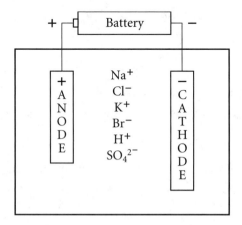

a.   The anode will attract ions of what charge? _____

b.   The cathode will attract ions of what charge? _____

a. −
b. +

121

a.   Ions attracted toward an anode are called *anions*. Anions have a _____ charge.

b.   Ions attracted toward a cathode are called *cations*. Cations have a _____ charge.

a. −
b. +

122

Which are the cations in the solution indicated in problem 120?

_____

Na⁺, K⁺, and H⁺

123

Which are the anions in that same solution? _____

Cl^-, Br^-, and SO_4^{2-}

124

a. The principal cation inside animal cells (the principal *intra*cellular cation), the potassium ion, has a _____ (positive/negative) charge.

b. The principal cation outside animal cells (the principal *extra*cellular cation), the sodium ion, has a _____ (positive/negative) charge.

c. The iodide ion, I^-, is necessary for the proper functioning of the thyroid gland. The iodide ion is a(n) _____ (anion/cation).

a. **positive**
b. **positive**
c. **anion**

pH

Mathematically, pH $= -\log H^+$. A log, or logarithm, is an exponent. The log (exponent) of 10^{-3} is -3.

125

The log of 10^{-7} is _____.

-7

126

If $H^+ = 10^{-4}$, then pH $= -\log H^+ = -(-4) =$ _____.

4

a. 12
b. 1

127

That is, pH is the negative log of the H^+.

a.   If $H^+ = 10^{-12}$, pH = _____.

b.   If $H^+ = 10^{-1}$, pH = _____.

a. acid
b. neutral
c. acid

128

The acid or basic strength of a solution may be expressed in terms of a number called the *pH* of that solution. The pH scale expresses the concentration of hydrogen ions (and hydroxide ions) in solution.

The pH range is from 0 to 14, with a pH of 7 indicating a neutral solution. A pH below 7 indicates an *acid* solution.

a.   A pH of 3 indicates a(n) _____ solution.

b.   A pH of 7 indicates a(n) _____ solution.

c.   A pH of 6 indicates a(n) _____ solution.

a. 7
b. neutral

129

Water is very slightly ionized: $H_2O \longrightarrow H^+ + OH^-$. The hydrogen ion concentration (H^+) in water is 10^{-7}.

a.   What is the pH of water? _____

b.   Is water acid, basic, or neutral? _____

Although all pH values below 7 indicate acid solutions, there is a definite progression of acid strengths according to pH values.

A pH between 5 and 7 indicates a weak acid solution, between 2 and 5 indicates a moderately strong acid solution, and between 0 and 2 indicates a strong acid solution, as shown in the following chart:

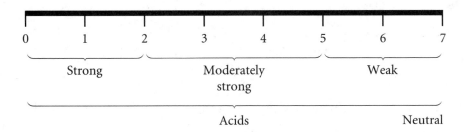

| 0 | 1 | 2 | 3 | 4 | 5 | 6 | 7 |

Strong   Moderately   Weak
                     strong

                          Acids   Neutral

130

a. Which of the following pH values indicates a solution containing a strong acid?

__ 4   __ 7   __ 1   __ 6

b. Which of the following pH values indicates a solution containing a moderately strong acid?

__ 4   __ 7   __ 1   __ 6

c. A pH of 0 indicates what strength acid solution? _____

d. A pH of 3 indicates what strength acid solution? _____

a. 1
b. 4
c. strong
d. moderately strong

131

pH values may be indicated as decimal values as well as whole numbers. Thus, a pH of 2.56 indicates a solution whose pH lies between 2 and 3. Thus, this solution is a:

__ strong acid solution
__ weak acid solution
__ moderately strong acid solution

moderately strong
acid solution

6.27

132

Among the following pH values, which solution contains a weak acid?

__ 1.72   __ 3.75   __ 2.00   __ 6.27

1.72

133

Among the following pH values, which solution contains a strong acid?

__ 1.72   __ 3.75   __ 7.00   __ 5.00   __ 6.38

7.00

134

Of solutions with the following pH values, which one is neutral?

__ 2.70   __ 4.65   __ 5.00   __ 7.00

A solution whose pH is above 7 is called a *basic* solution.

A solution whose pH is between 7 and 9 is called a weak basic solution, between 9 and 12 a moderately strong basic solution, and between 12 and 14 a strong basic solution, as shown in the following chart:

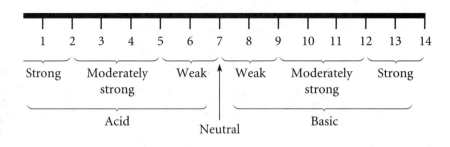

135

Of the following pH values, which indicates a strong basic solution?

__ 3.5   __ 7.0   __ 9.5   __ 13.9   __ 6.4

136

Of the following pH values, which indicates a weak basic solution?

__ 2.7   __ 8.3   __ 4.2   __ 10.8   __ 14.0

137

Which of the following pH values indicates a moderately strong basic solution?

__ 2.2   __ 4.7   __ 11.1   __ 13.7

138

The pH of saliva lies between 5.5 and 6.9, indicating a _____ solution.

139

Bile has a pH range of 7.8 to 8.6, so it is a _____ solution.

13.9

8.3

11.1

weak acid

weak basic

weak acid

strong acid

weak basic

weak basic

a. 10
b. weak

140

Urine has a pH range of 5.5 to 6.9, so it is a _____ solution.

141

The gastric juices have a pH range of 1.6 to 1.8, so they make up a _____ solution.

142

Blood has a pH range of 7.35 to 7.45, so it is a _____ solution.

143

The pancreatic juices have a pH range of 7.5 to 8.0, so they make up a _____ solution.

144

A difference of 1 in pH is equivalent to a 10-fold difference in acid or base strength. That is, an acid of pH 2 is 10 times as strong as an acid of pH 3. Likewise, a base of pH 10 is 10 times as strong as a base of pH 9.

a. A solution of pH 4.73 is _____ times as strong as one of pH 5.73.

b. A solution of pH 9.81 is 10 times as _____ (strong/weak) as one of pH 10.81.

145

Thus, a small difference in pH can represent a _____
(considerable/negligible) difference in acid or base strength.

considerable

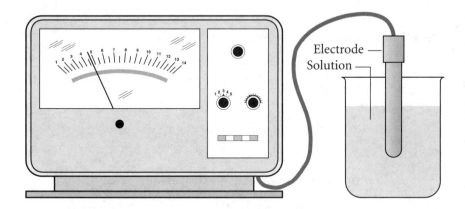

Electrode
Solution

BUFFERS

146

Recall that acids yield hydrogen ions (H^+) or protons in solution.
Both hydrochloric acid (HCl) and acetic acid (CH_3COOH) are acids,
as indicated by the following reactions:

$$HCl \longrightarrow H^+ + Cl^-$$
$$CH_3COOH \rightleftharpoons CH_3COO^- + H^+$$

The HCl, however, is 100% ionized, as shown by the one-way arrow,
whereas CH_3COOH is less than 1% ionized, as shown by the double
(equilibrium) arrows. That is, relatively speaking, HCl yields many
H^+ ions in solution, whereas CH_3COOH yields few H^+ ions.

a. few
b. strong
c. weak

Hydrochloric acid, HCl, is called a *strong acid*. It yields relatively many H^+ ions in solution.

a.   Acetic acid, CH_3COOH, is called a *weak acid*. It yields relatively _____ (many/few) H^+ ions in solution.

b.   Sulfuric acid, H_2SO_4, yields relatively many H^+ ions in solution. It is a _____ (strong/weak) acid.

c.   Citric acid, which yields relatively few H^+ ions in solution, is a _____ (strong/weak) acid.

147

a. many
b. few

A base accepts or reacts with protons (hydrogen ions).

a.   A strong base reacts with relatively _____ (many/few) H^+ ions.

b.   A weak base reacts with relatively _____ (many/few) H^+ ions.

148

a. strong
b. weak

a.   NaOH, sodium hydroxide, reacts with relatively many H^+ ions. It is a _____ (strong/weak) base.

b.   Sodium bicarbonate, $NaHCO_3$, reacts with relatively few H^+ ions. It is a _____ (strong/weak) base.

149

a. When a strong acid is added to water, the pH of the solution should go _____ (up/down).

b. When a strong base is added to water, the pH of the solution should go _____ (up/down).

The addition of a weak acid or a weak base to water should have very little effect on the pH.

a. down
b. up

150

A *buffer solution* is one that maintains a constant pH upon the addition of small amounts of either acid or base. A buffer consists of a weak acid and a weak base. A buffer solution can "sponge up" excess H^+ if added to a solution, or it can release H^+ if the H^+ concentration drops. One such buffer in the body is the $H_2CO_3/NaHCO_3$ system.

Suppose that a strong acid such as HCl is added to the buffer. The reaction is

$$HCl \ + \ NaHCO_3 \ \longrightarrow \ H_2CO_3 \ + \ NaCl$$
Strong acid   Weak base   Weak acid   Salt

Thus, the strong acid has been changed to a weak acid. Should the pH change? _____

no

151

If a strong base such as sodium hydroxide, NaOH, is added to the buffer, then the reaction is

$$NaOH \ + \ H_2CO_3 \ \longrightarrow \ NaHCO_3 \ + \ H_2O$$
Strong base   Weak acid   Weak base   Water

The strong base has been changed to a _____ base. Should the pH change? _____

weak; no

it should not change it

no

no effect

yes

152

Another buffer system in the body is the phosphate buffer system, which consists of NaH_2PO_4/Na_2HPO_4. What effect will the addition of a strong acid have on the pH of this buffer? _____

153

Should the addition of base affect the pH of a phosphate buffer system? _____

154

What effect should the removal of acid (which is equivalent in effect to the addition of base) have on the pH of a buffer solution? _____

155

The pH of the blood (a buffer solution) lies in the range of 7.35 to 7.45.

Normal metabolic processes add acid to the bloodstream. What effect should these acids have on the pH of the blood? _____.

156

Protein buffers are also present in the body. Should they behave similarly to the bicarbonate and phosphate buffers? _____

Liquid Mixtures 5

SOLUTIONS

A solution consists of a *solute* (a relatively less abundant substance) dissolved in a *solvent* (a relatively more abundant substance). In living organisms, the solvent is water.

157

a. When oxygen dissolves in water, oxygen is the _____ (solute/solvent).

b. When alcohol is dissolved in water, alcohol is the _____ (solute/solvent).

c. When salt is dissolved in water, water is the _____ (solute/solvent).

158

A solution consists of a _____ and a _____.

solution

159

Solute + solvent $\longrightarrow$ _____.

a. yes
b. homogeneous
c. transparent

160

Solutions are transparent and homogeneous. *Homogeneous* means "the same throughout," as compared with *heterogeneous*, which means "not the same throughout."

a. In a saltwater solution, is one part the same as any other part?

_____

b. Solutions are _____ (homogeneous/heterogeneous).

c. Solutions are _____ (transparent/opaque).

a. yes
b. yes

161

In general, solutions pass through membranes.

a. If glucose is administered intravenously, will the glucose solution pass through the membranes in the body? _____

b. Will saline (salt) solution pass through a membrane? _____

water

162

For medical use, solutions are frequently described in terms of *percent* or *parts per hundred*.

In a percent solution, the percentage number indicates the number of grams of solute present in 100 mL of solution. The solvent is usually _____.

163

a. A 5% glucose solution contains 5 g of glucose in _____ mL of solution.

b. A 2% boric acid solution contains _____ g of boric acid in _____ mL of solution.

164

Physiologic saline solution, 0.95% NaCl solution, contains _____ g of NaCl in _____.

165

Another method of expressing solution concentration is in terms of *molarity*, or *moles* of solute per liter of solution. A *mole* of a compound is the number of grams of that compound equal to its molecular mass.

The atomic mass of hydrogen is 1 and that of oxygen is 16. The molecular mass of water, H_2O, which consists of 2H and 1O, is $(2 \times 1) + (1 \times 16)$, or _____. Therefore, one mole of water has a mass of _____ g.

166

What is the molecular mass of glucose, $C_6H_{12}O_6$? _____
Atomic masses are C = 12, H = 1, O = 16.

a. 100
b. 2; 100

0.95; 100 mL of solution

18; 18

180 g

a. 1; 1 liter
b. 120; 1 liter
c. 0.1; solution
d. 5.85

167

A 1-molar (1M) glucose solution contains 1 mole (180 g) of glucose dissolved in enough water to make 1 liter of solution.

Use the following atomic masses in making calculations:

$$
\begin{array}{llll}
H & 1 & O & 16 \\
Mg & 24 & S & 32 \\
Na & 23 & Cl & 35.5
\end{array}
$$

a. To prepare a 1M $MgSO_4$ solution, take _____ mole(s) of $MgSO_4$ and dissolve in enough water to prepare _____ (how much?) of solution.

b. To prepare a 1M $MgSO_4$ solution, take _____ g of $MgSO_4$ and dissolve in enough water to make _____ (how much?) of solution.

c. To prepare a 0.1M NaCl solution, take _____ mole(s) of NaCl and dissolve in enough water to make 1 liter of _____.

d. To prepare a 0.1M NaCl solution, take _____ g of NaCl and dissolve in enough water to make 1 liter of solution.

168

Osmolarity is a method of expressing solution concentration based on the number of particles in solution.

$$
\text{Osmolarity} = \text{molarity} \times \frac{\text{number of particles}}{\text{molecule}}
$$

Osmolarity is expressed in the units osmoles, or *osmol*.

Consider the following reactions:

$$NaCl \longrightarrow Na^+ + Cl^-$$
1 molecule $\longrightarrow$ 2 particles

$$Glucose \longrightarrow glucose$$
1 molecule $\longrightarrow$ 1 particle

$$K_2SO_4 \longrightarrow 2K^+ + SO_4^{2-}$$
1 molecule $\longrightarrow$ 3 particles

a.   A 1-molar (1M) glucose solution has a concentration of _____ osmol.

b.   A 0.5M NaCl solution has a concentration of _____ osmol.

c.   A 0.4M K_2SO_4 solution has a concentration of _____ osmol.

a.   1
b.   1
c.   1.2

169

Osmolarity may also be expressed in the units milliosmoles, or *mosmol*, where

$$1 \text{ osmol} = 1000 \text{ mosmol}$$

To change osmoles to milliosmoles, multiply by 1000.

a.   A 0.15M NaCl solution has a concentration of _____ osmol.

b.   A 0.15M NaCl solution has a concentration of _____ mosmol.

Note that human body fluids have an osmolarity of approximately 300 milliosmol.

a.   0.30
b.   300

SUSPENSIONS

170

When sand is placed in water and shaken, a sand-water suspension is produced.

a.   When the shaking is stopped, does the sand settle? _____

b.   Does sand dissolve in water? _____

c.   In a suspension, is there a solute? _____ Why or why not?

_____

Note that particles in a suspension are so large that they settle under the influence of gravity.

a.   yes
b.   no
c.   no; the substance does not dissolve in water

171

a.   A suspension is _____ (homogeneous/heterogeneous).

b.   A solution is _____ (homogeneous/heterogeneous).

c.   Is a suspension transparent? _____

d.   Is a solution transparent? _____

a.   heterogeneous
b.   homogeneous
c.   no
d.   yes

172

Blood is a suspension of blood cells in blood plasma. If blood is treated so that it doesn't clot and is then placed in a test tube, will the blood cells settle? _____

What causes the blood cells to settle? _____

yes; gravity

COLLOIDS
(COLLOIDAL DISPERSIONS)

Colloids consist of very tiny particles suspended in a liquid, usually water. (Colloids may also be suspended in solids and in gases.) Colloids differ from suspensions in that colloids do not settle. Colloidal dispersions are usually translucent or milky white. They do not pass through membranes.

173

One example of a colloid is protein.

a. A colloidal dispersion of protein will be _____ (transparent/ translucent).

b. A colloidal dispersion of protein _____ (will/will not) settle.

c. A colloidal dispersion of protein _____ (will/will not) pass through a membrane.

174

Many proteins are colloids. Should such proteins be able to pass through the membranes of the kidneys and be present in the urine? _____

Should salt, NaCl? _____

a. **translucent**
b. **will not**
c. **will not**

no; yes

Diffusion and Osmosis 6

DIFFUSION

If a bottle of perfume is opened in a room, the odor is soon apparent throughout the room. The molecules of perfume diffuse, or spread out, into the available volume of air.

Next, suppose that a sugar solution is placed on one side of a permeable (nonselective) membrane and that pure water is placed on the other side of that membrane.

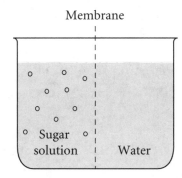

The sugar molecules (the solute) will diffuse across the permeable membrane until the concentrations on both sides of the membrane are equal.

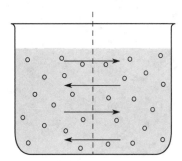

When this equilibrium occurs, sugar (solute) molecules will continue to diffuse across the membrane in both directions so that there will be no net change in concentration.

Diffusion is the tendency to spread into available space. That is, a substance will diffuse from where it is more concentrated to where it is less concentrated. Diffusion across a membrane is called *passive transport* because it requires no external energy.

175

During diffusion, solute particles move from:

__ high concentration of solute to lower concentration of solute
__ low concentration of solute to higher concentration of solute

high concentration of solute to lower concentration of solute

176

In passive transport, solute particles flow from a _____ (high/low) concentration to one of _____ (higher/lower) concentration.

high; lower

177

Can diffusion occur across a permeable membrane? _____

yes

178

Can diffusion occur without a membrane? _____

yes

179

Do all solutions have the same solute concentration? _____

OSMOSIS

A solution that has a higher solute concentration than another solution is said to be *hypertonic* (hyperosmotic) when compared with the second solution.

A solution that has a lower solute concentration than another solution is said to be *hypotonic* (hypoosmotic) when compared with the second solution.

A solution that has the same solute concentration as another solution is said to be *isotonic* (isoosmotic) when compared with the second solution.

Look at the following illustrations of solutions whose solute is indicated by small circles.

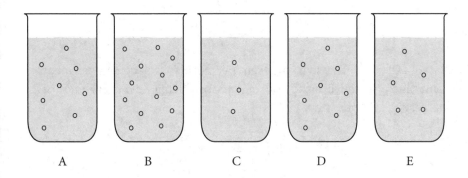

A   B   C   D   E

180

a.   Solution A is hypertonic to solution(s) _____.

b.   Solution A is hypotonic to solution(s) _____.

c.   Solution A is isotonic to solution(s) _____.

no

a. C, E
b. B
c. D

nonselective

181

A permeable membrane is _____ (selective/nonselective).

A semipermeable membrane is one that is selectively permeable.

Consider the following system in which a selectively permeable membrane separates a dilute (weak) sugar solution from a concentrated (strong) sugar solution. We will assume that this semipermeable membrane is permeable to water but not to sugar molecules.

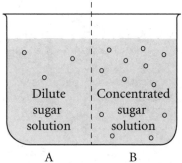

Water will flow through the semipermeable (selectively permeable) membrane and equalize the concentrations. Such a flow is called *osmosis*. The results will be:

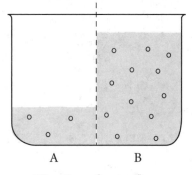

182

Osmosis is a type of diffusion. Does it require energy? _____
Is osmosis passive transport? _____

no; yes

183

Osmosis is the flow of _____ across a membrane.

water

184

Original solution A is _____ (hypertonic/hypotonic) when
compared with original solution B.

hypotonic

185

Original solution B is _____ when compared with original
solution A.

hypertonic

186

During osmosis, water flows from a hypotonic solution to a hypertonic
solution, or from a _____ (weaker/stronger) solution to a
_____ (weaker/stronger) solution.

weaker; stronger

187

Osmosis is the flow of _____ through a selectively permeable
membrane.

water

yes

188

Will osmosis continue until the two solutions have equal concentrations?
_____

solute; water molecules

189

Diffusion of sugar is the flow of _____ particles; osmosis is the flow of _____ through a selectively permeable (semipermeable) membrane.

hypertonic (concentrated); hypotonic (dilute)

190

Diffusion is the flow of solute particles from a _____ solution to a _____ solution.

hypotonic (dilute); hypertonic (concentrated)

191

Osmosis is the flow of water through a membrane from a _____ solution to a _____ solution.

yes; yes

192

Is diffusion passive transport? _____ Is osmosis? _____

193

a. hypotonic
b. hypertonic
c. hypertonic
    (concentrated);
    hypotonic (dilute)

Consider the following system where dilute and concentrated salt solutions are separated by a selectively permeable membrane that allows the salt ions and water to flow through.

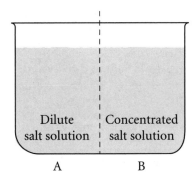

a. What term is used to compare the strength of solution A with that of solution B? _____

b. What term is used to compare the strength of solution B with that of solution A? _____

c. When diffusion takes place, the flow of solute will be from _____ to _____ solution.

194

passive

Such a flow of solute is also called _____ transport.

195

water; hypotonic;
hypertonic

While diffusion is taking place, osmosis can also occur.

During osmosis, _____ flows through a membrane from a _____ solution to a _____ solution.

no (opposite directions)

Selectively permeable membrane

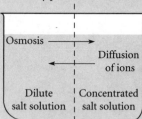

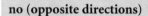

Does osmosis take place in the same direction through the membrane as diffusion of a solute? _____

197

In the following diagram, indicate the direction of flow of osmosis and diffusion of ions.

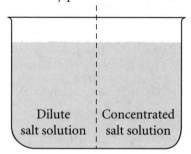

Selectively permeable membrane

OSMOTIC PRESSURE

198

Consider the following system in which water is placed on both sides of a U-tube that has a selectively permeable membrane (permeable to water but not to solutes) dividing the two halves.

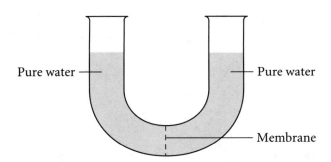

Will osmosis take place? _____ Why or why not? _____

_____

199

Next, consider a similar U-tube in which water is placed on one side and sugar solution on the other. The membrane is impermeable to sugar but permeable to water.

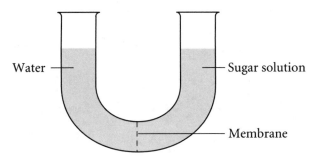

a.   Will osmosis take place? _____

b.   In which direction will osmosis take place? _____

The net result will be a rise in the liquid level on the sugar side and a drop in the liquid level on the water side, as in the following figure.

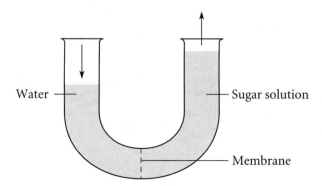

If a piston is placed on the sugar side of the U-tube and pressure is applied to that piston, then osmosis can be prevented. The amount of pressure required to counteract osmosis is called the *osmotic pressure*.

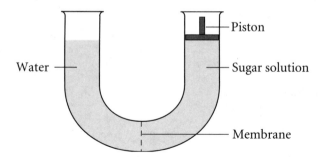

In general, the greater the difference in concentration on two sides of a membrane, the greater the osmotic pressure.

C; B

200

In the following diagrams of sugar solutions separated from each other by a membrane impermeable to sugar but permeable to water, which will exert the greatest osmotic pressure? _____ The least? _____

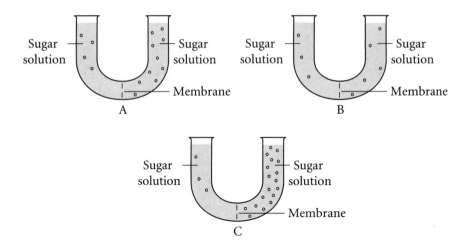

201

Consider the following system where a U-tube has water on one side of a membrane and a dilute NaCl solution on the other. The membrane is permeable to sodium and chloride ions and to water.

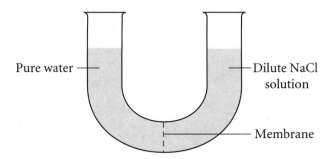

a. Will osmosis take place? _____

b. In which direction? _____

c. Will diffusion take place? _____

d. In which direction? _____

e. Will the two concentrations equalize? _____

a. yes
b. into NaCl solution
c. yes
d. into the water
e. yes

a. yes
b. yes
c. yes
d. yes
e. diffusion, or passive transport

202

Now consider a U-tube with NaCl solution on one side and KCl solution on the other; the solutions have the same concentrations. The membrane is equally permeable to both NaCl and KCl.

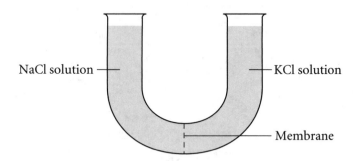

a.   Will NaCl diffuse? _____

b.   Will KCl diffuse? _____

c.   Will the concentrations of KCl be the same on each side when diffusion is finished? _____

d.   Will the concentrations of NaCl be the same on each side?

_____

e.   What is this motion of solute particles called? _____

ACTIVE TRANSPORT

203

Next, consider an animal cell with cation concentrations in millimoles per liter as listed in the following table.

Cation	Intracellular	Extracellular
Na	10	140
K	140	4

yes

The same situation is shown in the following diagram, where the size of the symbol indicates relative concentration.

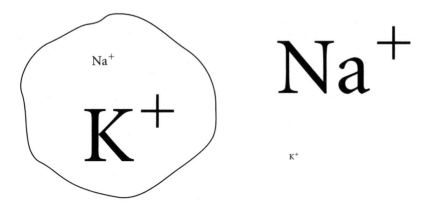

If diffusion or passive transport were the only factor affecting these ions, would the concentrations tend to equalize? _____

$$\boxed{204}$$

Because the extracellular and intracellular concentrations of sodium and potassium ions are unequal in animal cells, there must be another force acting (shown in the following diagram).

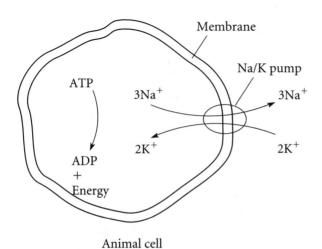

Animal cell

a. ATP
b. Na^+
c. K^+
d. no
e. 2

This type of movement is called *active transport.*

a.   What energy compound is required for active transport?

_____

b.   What ion does active transport pump out of the cell? _____

c.   What ion does active transport pump into the cell? _____

d.   Are the movements of Na^+ and K^+ equal in amount? _____

e.   For every three sodium ions pumped out of a cell, how many potassium ions are pumped in? _____

205

This type of ion movement is called _____ transport and requires the energy compound _____.

206

This ion system pumping sodium ions out of a cell and potassium ions into a cell is called a *sodium-potassium pump.*

Are Na^+ and K^+ pumped equally across a membrane? _____

207

How many Na^+ are pumped out? How many K^+ are pumped in?

_____

208

Because both Na^+ and K^+ are positively charged ions, 3 positive charges are pumped out of a cell for every 2 that are pumped in. Should the inside of the cell be more positive or less positive than the outside of the cell? _____

active; ATP

no

3; 2

less

209

Should the inside of the cell be positive or negative when compared with the outside of the cell? _____

210

Animal cells have a *membrane potential* or a small voltage difference between the inside and outside of the cell.

An animal cell has a small voltage difference or _____ between the inside and outside of the cell.

211

Complete the following diagram.

_____ transport

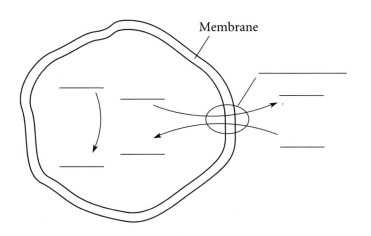

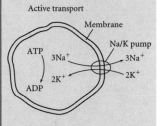

Nerve Cells 7

DEPOLARIZATION AND REPOLARIZATION

212

A resting neuron has a higher _____ (K^+/Na^+) concentration inside the cell and a higher _____ (K^+/Na^+) ion concentration outside that cell.

K^+; Na^+

213

A resting cell has a $(+/-)$ charge inside.

$-$

214

A resting neuron has a fairly constant electrical potential difference (voltage) of -50 to -100 millivolts inside. The following diagram represents a resting neuron. Note that the membrane potential is between -80 and -100 mV. Does a resting neuron have a positive or negative electrical potential inside? _____

negative

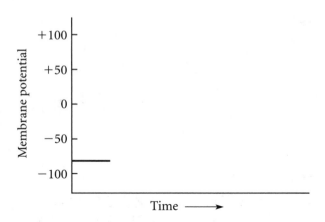

a. **closed potassium channel**

b. **open potassium channel**

215

Some cells, particularly nerve and muscle cells, can generate changes in the electrical potential difference across their cell membranes when a stimulus is applied. Such cells have sodium and potassium channels that can open to allow these ions in or out of the cell. When closed, these channels do not allow ions to pass through.

A sodium channel may be represented by a square shape (S for sodium; S for square). The diagram on the left represents an open sodium channel, and the diagram on the right represents a closed sodium channel.

If a triangular shape represents a potassium channel, what do the following diagrams represent? a. _____ b. _____

a   b

216

The following diagram illustrates a resting neuron. Which channels are open? _____ Closed? _____

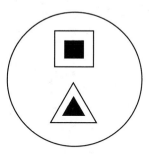

both Na and K channels are closed

217

If both Na and K channels are closed, should the intracellular Na^+ and K^+ concentrations remain constant?

yes

218

If a neuron is stimulated, then the sodium channels open but the potassium channels remain closed, as in the following diagram.

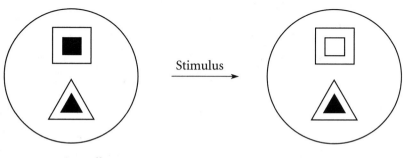

Resting cell

a. Will sodium ions flow into or out of the cell? _____

b. Will potassium ions flow into or out of the cell? _____

a. into
b. neither (the potassium channel is closed)

+

a stimulus

resting state

219

Sodium ions have what type of charge, + or −? _____

As sodium ions flow into the cell, the potential of that cell will change from − to +. The cell is said to be *depolarized*.

220

What causes the sodium channels to open? _____

221

As the cell becomes depolarized, the membrane potential changes from − to +. Section 2 of the following diagram illustrates depolarization. What does section 1 indicate? _____

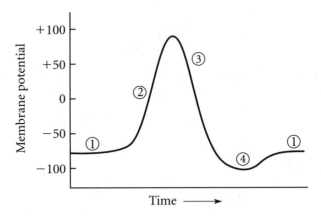

222

Color in the following diagram to show which channels should be open and which should be closed. What is effect *A*?

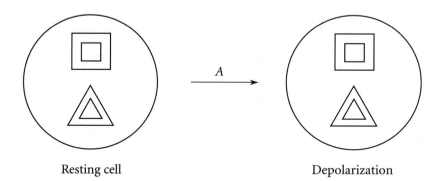

Resting cell   Depolarization

223

Once the cell has become depolarized, the sodium channels close and the potassium channels open.

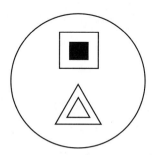

Will sodium ions flow into or out of the cell? _____

Will potassium ions flow into or out of the cell? _____

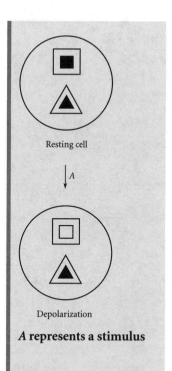

Resting cell

↓*A*

Depolarization

A represents a stimulus

neither; out

it becomes negative

224

This phase is called *repolarization*, as shown in section 3 of the following diagram.

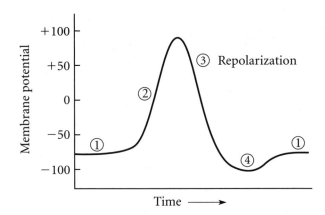

During repolarization, as potassium ions flow out of the cell, what happens to the membrane potential? _____

resting state; depolarization

225

What is section 1 called in the problem 224 diagram? _____

What is section 2 called? _____

226

Color the following diagram to indicate which of the following channels are open or closed.

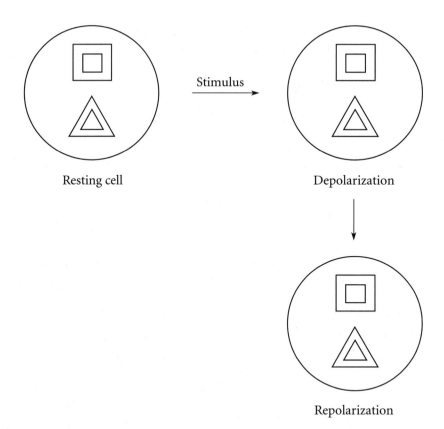

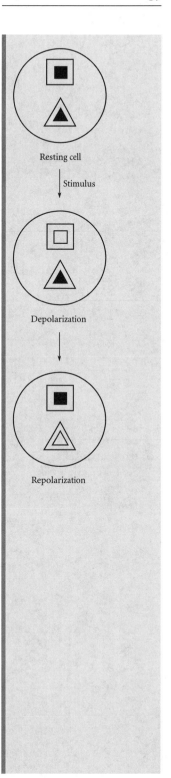

Resting cell

Stimulus

Depolarization

Repolarization

The flow of potassium ions out of the cell during repolarization continues afterward because these channels are slow to close. Such continued flow produces an *undershoot* or greater negative charge inside the cell than was present in the resting state, as shown in the following diagram.

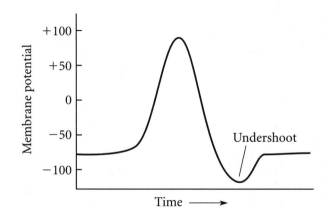

Finally, both channels are closed and active transport takes over to return the cell to its resting state.

227

a. sodium
b. potassium
c. $3 Na^+$ out; $2 K^+$ in
d. potassium
e. sodium

a.   During active transport, _____ ions are carried out of the cell.

b.   During active transport, _____ ions are carried into the cell.

c.   What are the relative number of Na^+ and K^+ ions moved out of and into the cell? _____

d.   During undershoot, which channels are slow to close?

_____

e.   During undershoot, which channels remain closed? _____

228

1: resting state
2: depolarization
3: repolarization
4: undershoot
5: resting state

In the following diagram, label sections 1 through 5.

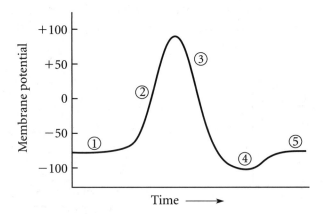

229

In the following diagram, indicate the charges inside the cell membrane. Also mark which channels are open and which are closed. Finally, label *A* and *B*.

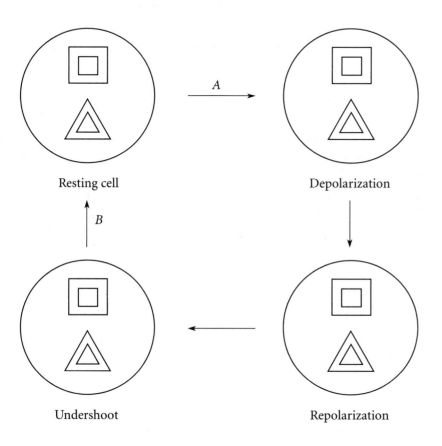

Resting cell

A

Depolarization

B

Undershoot

Repolarization

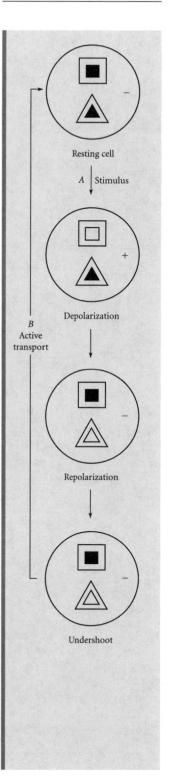

Resting cell

A | Stimulus

Depolarization

B
Active
transport

Repolarization

Undershoot

PROPAGATION OF A NERVE IMPULSE

Nerve impulses are propagated in an axon by means of a series of depolarizations and repolarizations, such as indicated in the following drawing.

+ + + +	– – – –	+ + + +
– – – –	+ + + +	– – – –
– – – –	+ + + +	– – – –
+ + + +	– – – –	+ + + +
Resting state	Depolarization	Repolarization

How does this propagation occur? Depolarization increases the membrane potential to a critical level called the threshold potential. This increase in turn triggers an *action potential*, or nerve impulse, which is transmitted to an adjacent segment. Consider the following diagram of the cross section of a small section of an axon that illustrates three adjacent segments in their resting state with a negative potential inside each segment.

+	+	+	+	+	+	+	+	+
–	–	–	–	–	–	–	–	–
–	–	–	–	–	–	–	–	–
+	+	+	+	+	+	+	+	+

Segment A   Segment B   Segment C

When a stimulus is applied to segment A, it causes that segment to become depolarized or positively charged in its inner membrane, which produces an action potential.

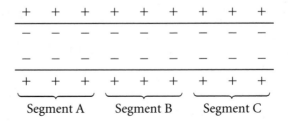

–	–	–	+	+	+	+	+	+
+	+	+	–	–	–	–	–	–
+	+	+	–	–	–	–	–	–
–	–	–	+	+	+	+	+	+

Segment A   Segment B   Segment C

Depolarized   Resting state   Resting state

The depolarization of segment A in turn causes the depolarization of segment B so that the action potential (nerve impulse) travels along the indicated path.

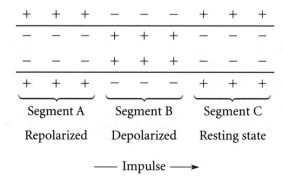

	+ + +	− − −	+ + +
	− − −	+ + +	− − −
	− − −	+ + +	− − −
	+ + +	− − −	+ + +

Segment A   Segment B   Segment C

Repolarized   Depolarized   Resting state

—— Impulse ⟶

230

While the impulse is traveling from A to B, segment A becomes repolarized and returns to its resting state.

The depolarization of segment B will have what effect on segment C?

_____

it will become
depolarized

231

What happens to the action potential (nerve impulse)? _____

_____

it travels from segment B
to segment C

232

What happens to segment B after the impulse is transmitted to C? _____

_____

it becomes repolarized
and then returns to its
resting state

This effect may be considered to be similar to that of a section of standing dominoes. When the first one is pushed over, the motion is transmitted to succeeding ones just as a nerve impulse travels from one end of an axon to the other.

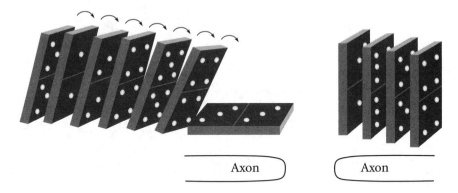

Axon   Axon

Note, however, that the last falling domino is too far from the next one to cause it to fall. The effect then stops.

Similarly, the movement of a nerve impulse travels along an axon to the end but cannot jump across the gap to the next axon. Something must carry that impulse across the gap.

233

The following diagram illustrates the end of one axon and the beginning of the succeeding one.

Can the nerve impulse jump across the synaptic cleft? _____

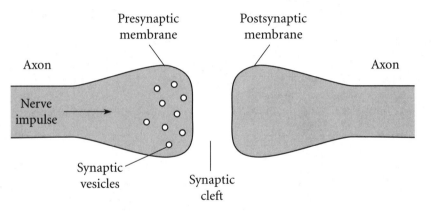

Presynaptic membrane   Postsynaptic membrane

Axon   Axon

Nerve impulse →

Synaptic vesicles

Synaptic cleft

no

234

When the action potential reaches the presynaptic membrane, it depolarizes it. The depolarization of the presynaptic membrane triggers an influx of calcium ions, which in turn causes the synaptic vesicles to fuse with the presynaptic membrane.

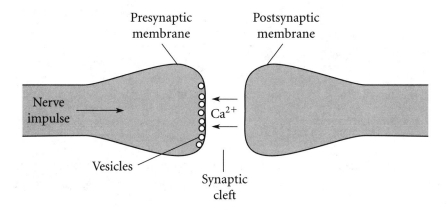

What effect does a nerve impulse have on the presynaptic membrane?

_____

235

The depolarization of the presynaptic membrane causes the influx of which ions? _____

236

The influx of calcium ions has what effect on the vesicles?

_____

it depolarizes it

calcium

they fuse with the pre-synaptic membrane

When the vesicles fuse to the presynaptic membrane, they release neurotransmitters that can travel across the synaptic cleft to the postsynaptic membrane.

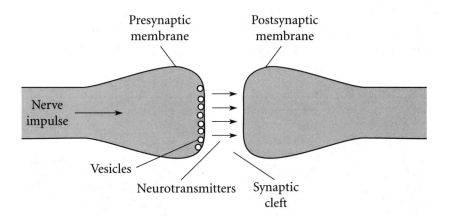

At the postsynaptic membrane, the neurotransmitters bind to receptor sites, which then open specific ion channels in the postsynaptic membrane.

yes

237

When an ion channel is open, can ions move across a membrane?

_____

no

238

When an ion channel is closed, can ions move across a membrane?

_____

neurotransmitter

239

When the vesicles bind to the presynaptic membrane, what type of substance is released? _____

240

When a neurotransmitter is released, where does it go? _____

into the synaptic cleft

241

When the neurotransmitter reaches the postsynaptic membrane, what effect does it have on receptor sites? _____

it causes ion channels to open

Once the ion channels open (which channels depends on the the receptor molecules associated with those channels), selected ions enter the postsynaptic membrane, causing depolarization and thus a new action potential there. The new action potential then carries the nerve impulse along that axon.

242

Does the identity of the neurotransmitter determine which ion enters the postsynaptic membrane? _____

yes

243

When the ion channels are open, selected ions can enter the postsynaptic membrane, causing _____.

depolarization

244

When the postsynaptic membrane is depolarized, what effect may be produced? _____

an action potential

it begins sending a nerve impulse along that axon

245

What effect does this action potential have on the second axon?

After the postsynaptic membrane is depolarized, enzymes degrade the neurotransmitters so that the membrane is ready for succeeding signals from new molecules of neurotransmitters.

A: presynaptic membrane
B: synaptic cleft
C: postsynaptic membrane
D: vesicles
E: ion channels
F: action potential
G: flow of neurotransmitters

246

Label *A* through *G* in the following diagram.

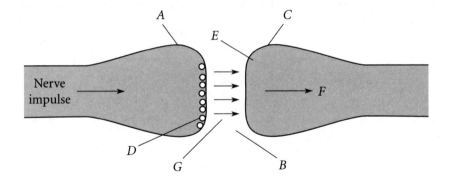

PART **II**

Organic
Chemistry

The Covalent Bond 8

247

You've already learned that atoms tend to lose electrons to form positively charged ions if they have 1, _____, or _____ electrons in their valence shells.

On the other hand, atoms that have _____ or _____ electrons in their valence shells tend to gain electrons to form _____ charged ions.

2; 3; 6; 7; negatively

248

Atoms with 4 or 5 electrons in their valence shells tend to *share* electrons. When atoms share electrons, they form a *covalent bond*.

Which of the following illustrated atoms would form covalent bonds?

_____

C and N

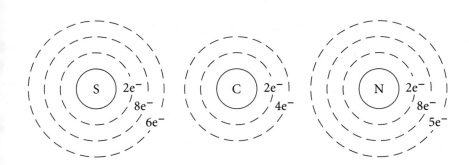

a. Mg, H, Al
b. Cl, S
c. Ne
d. stable and unreactive

249

Answer the following questions about these illustrated atoms.

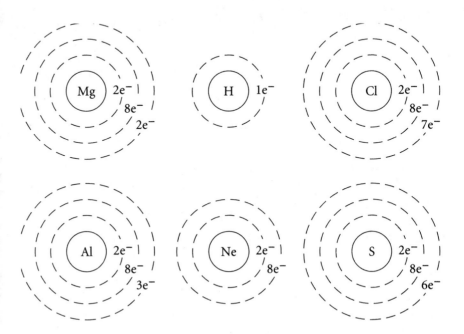

a. Which of the atoms will lose electron(s) to form positively charged ions? _____

b. Which of the atoms will gain electron(s) to form negatively charged ions? _____

c. Which of the atoms has a complete valence shell? _____

d. You know that elements with a complete valence shell are usually:

    __ unstable and reactive   __ stable and unreactive

250

a. covalent
b. nonelectrolyte

a.   When two atoms share electrons, they are held together by a
_____ bond.

b.   When two atoms share electrons, no ions are produced. Is the
resulting compound an electrolyte or a nonelectrolyte? _____

251

ionic

Atoms with 4 or 5 electrons in their valence shells tend to share electrons, but under some conditions atoms with more or fewer electrons
in their valence shells may also share electrons.

When one atom loses an electron and another atom gains that
electron, the ions thus formed are held together by a(n) _____
(covalent/ionic) bond.

252

a. electrolytes
b. nonelectrolytes

a.   Compounds containing ionic bonds are _____
(electrolytes/nonelectrolytes).

b.   Compounds containing only covalent bonds are _____
(electrolytes/nonelectrolytes).

4

253

The carbon atom, atomic number 6, has 4 electrons in its valence shell. It thus needs _____ more electrons to become stable. It can get these electrons by sharing. The 4 electrons in the valence shell of the carbon atom may be represented as here, with one dot for each electron in the valence shell.

$$\cdot \; C \; \cdot$$

2

254

When atoms share electrons, they always try to reach stable configurations of 8 electrons in the valence shells. The only exception to this rule is the hydrogen atom. It reaches a stable configuration when it has 2 electrons in its valence shell, which is also its only energy level. This first energy level can hold only _____ electrons.

4

255

Consider the carbon atom with 4 electrons in its valence shell. With how many hydrogen atoms can it share electrons to reach a stable configuration of 8 electrons in its valence shell? _____

256

The four hydrogen atoms can share electrons with a carbon atom to form a compound of the following type:

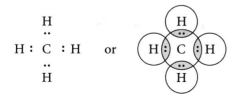

There are _____ electrons around the carbon atom and _____ electrons around each hydrogen atom.

257

When two hydrogen atoms combine, they share electrons. Each hydrogen atom then has _____ electrons around it.

258

In the following compound, what is the total number of covalent bonds? _____

$$
\begin{array}{c}
\text{H} \\
\text{··} \\
\text{H} : \text{C} : \text{H} \\
\text{··} \\
\text{H}
\end{array}
$$

8; 2

2

4

7

259

The chlorine atom, atomic number 17, has _____ electrons in its valence shell.

The electrons in the valence shell of the chlorine atom may be represented as

$$\cdot \; \overset{\textstyle ..}{\underset{\textstyle ..}{Cl}} \; :$$

When two chlorine atoms combine, they share electrons to reach a stable configuration of 8 electrons in the valence shell of each atom, or

covalent

260

What type of bond is there between two chlorine atoms? _____

261

A covalent bond is frequently indicated by a short line rather than by dots; thus, the hydrogen molecule may be represented as H : H or H — H. The short line indicates a pair of shared electrons.

The compound CH_4 may be represented as

$$
\begin{array}{ccc}
& H & \\
& \overset{\textstyle ..}{\underset{\textstyle ..}{H : C : H}} & \\
& H &
\end{array}
\qquad or \qquad
\begin{array}{ccc}
& H & \\
& | & \\
H & -C- & H \\
& | & \\
& H &
\end{array}
$$

Give the second representation of the chlorine molecule, Cl_2.

$$\overset{..}{:}\underset{..}{Cl}:\overset{..}{\underset{..}{Cl}}: \text{ or}$$

262

a. The carbon atom, atomic number 6, needs how many electrons to complete its valence shell? _____

b. Because the carbon atom has 4 electrons in its valence shell, and because it will tend to share these electrons, how many covalent bonds will a carbon atom form? _____

The carbon atom, therefore, must have four covalent bonds, or must have four bonds attached to it. These bonds may be indicated as follows:

$$-\overset{|}{\underset{|}{C}}- \quad -\overset{|}{C}= \quad -C\equiv \quad =C=$$

Note that each carbon atom has four bonds attached to it, regardless of how the bonds are arranged.

263

The hydrogen atom has 1 electron in its outer energy level, the first energy level. When the hydrogen atom shares electrons, how many more does it need to complete that valence shell? _____

Cl — Cl

a. 4
b. 4

1

a. 6
b. 2
c. 2

264

a. The oxygen atom, atomic number 8, has _____ electrons in its valence shell.

b. When the oxygen atom shares electrons, how many more electrons does it need to complete its valence shell? _____

c. Therefore, each oxygen atom must have how many bonds attached to it? _____

265

Draw the structure of the compound formed when four hydrogen atoms form bonds with a central carbon atom.

$$
\begin{array}{c}
\text{H} \\
| \\
\text{H}-\text{C}-\text{H} \\
| \\
\text{H}
\end{array}
$$

C

a. 3
b. 3

266

The arrangement of carbon atoms C — C indicates a bond or a shared electron pair between two carbon atoms.

a. How many hydrogen atoms may be attached to the left carbon atom? _____

b. How many hydrogen atoms may be attached to the right carbon atom? _____

267

a. Diagram all the hydrogen atoms attached to this structure:

C—C

b. How many bonds does each carbon atom have? _____

c. How many bonds does each hydrogen atom have? _____

a.

$$H-\underset{\underset{H}{|}}{\overset{\overset{H}{|}}{C}}-\underset{\underset{H}{|}}{\overset{\overset{H}{|}}{C}}-H$$

b. 4
c. 1

268

Consider this arrangement of carbon atoms:

C—C—C

a. How many hydrogen atoms may be attached to the left carbon atom? _____

b. How many hydrogen atoms may be attached to the center carbon atom? _____

c. How many hydrogen atoms may be attached to the right carbon atom? _____

a. 3
b. 2
c. 3

269

Diagram the compound containing three attached carbon atoms, indicating all the hydrogen atoms connected to those carbon atoms.

C—C—C

$$H-\underset{\underset{H}{|}}{\overset{\overset{H}{|}}{C}}-\underset{\underset{H}{|}}{\overset{\overset{H}{|}}{C}}-\underset{\underset{H}{|}}{\overset{\overset{H}{|}}{C}}-H$$

a. 3
b. 3
c. 3
d. 3
e. 0 (or none)

Consider this arrangement of carbon atoms:

a. How many hydrogen atoms may be attached to the right carbon atom? _____

b. How many hydrogen atoms may be attached to the left carbon atom? _____

c. How many hydrogen atoms may be attached to the upper carbon atom? _____

d. How many hydrogen atoms may be attached to the lower carbon atom? _____

e. How many hydrogen atoms may be attached to the center carbon atom? _____

a.

```
                H
                |
          H — C — H
      H         |         H
      |         |         |
  H — C ———— C ———— C — H
      |         |         |
      H         |         H
          H — C — H
                |
                H
```

b. it already has four bonds

a. Diagram the structure of the compound containing the following arrangement of carbon atoms, indicating all the hydrogen atoms.

```
          C
          |
      C — C — C
          |
          C
```

b. Why can there be no hydrogen atoms attached to the center carbon atom?_____

272

Consider this arrangement of carbon atoms:

$$C = C$$

There are two bonds (a double bond) between the carbon atoms. Each bond (a single bond) represents 1 pair of shared electrons. A double bond represents _____.

273

a. To become stable, a carbon atom must form how many covalent bonds? _____

b. In the structure $C = C$, how many hydrogen atoms may be attached to the left carbon atom? _____

c. In the space provided, diagram the structure of the compound containing two carbon atoms connected by a double bond. Show all the hydrogen atoms.

When you have completed your diagram, check it by counting the number of bonds around each carbon atom.

---

2 pairs of shared electrons

a. 4
b. 2
c.
$$\begin{array}{cc} H & H \\ | & | \\ H - C = C - H \end{array}$$

a. 2
b. 1
c. 3

274

Consider this arrangement of carbon atoms:

$$C = C - C$$

a. How many hydrogen atoms may be attached to the left carbon atom? _____

b. How many hydrogen atoms may be attached to the center carbon atom? _____

c. How many hydrogen atoms may be attached to the right carbon atom? _____

$$\begin{array}{ccc} H & H & H \\ | & | & | \\ H-C & = C - C - H \\ & & | \\ & & H \end{array}$$

275

Complete the structure of the following compound, showing all the hydrogen atoms.

$$C = C - C$$

$$\begin{array}{c} Cl \\ | \\ Cl-C-Cl \\ | \\ Cl \end{array}$$

276

Diagram the structure of carbon tetrachloride, CCl_4, in which each chlorine atom has one bond attached to it.

277

Diagram the structure of chloroform, CHCl₃.

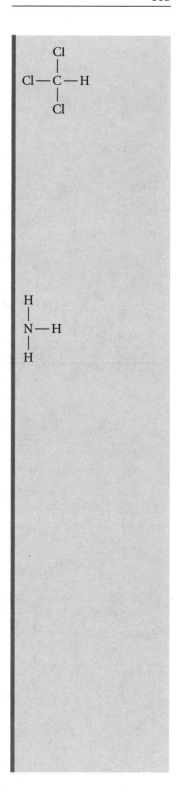

278

Nitrogen, atomic number 7, has 5 electrons in its valence shell and so must have three bonds attached to it. Diagram the ammonia molecule, NH₃.

Note that the structure

$$
\begin{array}{c}
\;\;\;\;\;\text{H}\;\;\;\;\;\text{H}\;\;\;\;\;\text{H}\;\;\;\;\;\text{H} \\
\;\;\;\;\;|\;\;\;\;\;\;|\;\;\;\;\;\;|\;\;\;\;\;\;| \\
\text{H}-\text{C}-\text{C}-\text{C}-\text{C}-\text{H} \\
\;\;\;\;\;|\;\;\;\;\;\;|\;\;\;\;\;\;|\;\;\;\;\;\;| \\
\;\;\;\;\;\text{H}\;\;\;\;\;\text{H}\;\;\;\;\;\text{H}\;\;\;\;\;\text{H}
\end{array}
$$

may be simplified and written as $CH_3CH_2CH_2CH_3$. Likewise,

$$
\begin{array}{c}
\;\;\;\;\;\text{H}\;\;\;\;\;\text{H}\;\;\;\;\;\text{H}\;\;\;\;\;\text{H}\;\;\;\;\;\text{H} \\
\;\;\;\;\;|\;\;\;\;\;\;|\;\;\;\;\;\;|\;\;\;\;\;\;|\;\;\;\;\;\;| \\
\text{H}-\text{C}-\text{C}-\text{C}-\text{C}-\text{C}-\text{H} \\
\;\;\;\;\;|\;\;\;\;\;\;|\;\;\;\;\;\;|\;\;\;\;\;\;|\;\;\;\;\;\;| \\
\;\;\;\;\;\text{H}\;\;\;\;\;\;|\;\;\;\;\;\text{H}\;\;\;\;\;\text{H}\;\;\;\;\;\text{H} \\
\;\;\;\;\;\;\;\;\;\;\;\text{H}-\text{C}-\text{H} \\
\;\;\;\;\;\;\;\;\;\;\;\;\;\;\;| \\
\;\;\;\;\;\;\;\;\;\;\;\;\;\;\;\text{H}
\end{array}
$$

may be simplified and written as $CH_3CH(CH_3)CH_2CH_2CH_3$, and

$$
\begin{array}{c}
\;\;\;\;\;\text{H}\;\;\;\;\;\text{H}\;\;\;\;\;\text{H}\;\;\;\;\;\text{H} \\
\;\;\;\;\;|\;\;\;\;\;\;|\;\;\;\;\;\;|\;\;\;\;\;\;| \\
\text{H}-\text{C}-\text{C}=\text{C}-\text{C}-\text{H} \\
\;\;\;\;\;|\;\;\;\;\;\;\;\;\;\;\;\;\;\;\;\;\;| \\
\;\;\;\;\;\text{H}\;\;\;\;\;\;\;\;\;\;\;\;\;\;\;\;\text{H}
\end{array}
$$

may be simplified and written as $CH_3CH{=}CHCH_3$.

279

How may the following structures be written in simplified form?

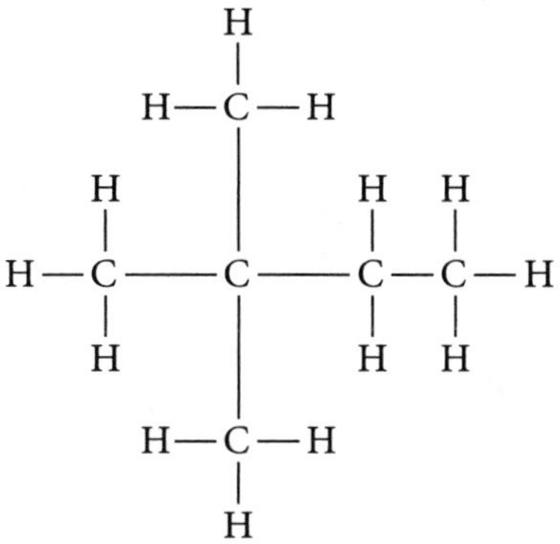

a. CH₃CH₃

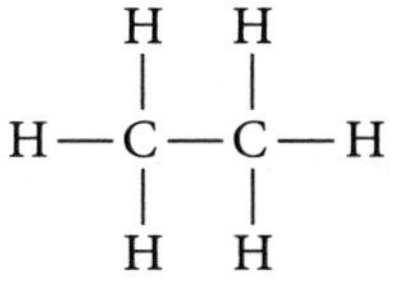

b. CH₃CH₂CH(CH₃)CH₂CH₃

c. CH₃C≡CH

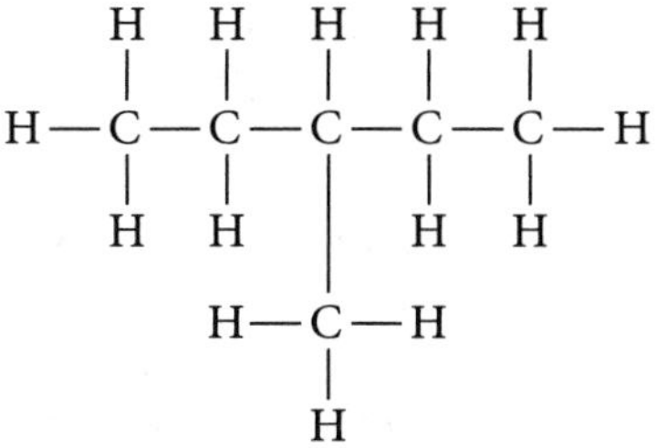

d. CH₃C(CH₃)₂CH₂CH₃

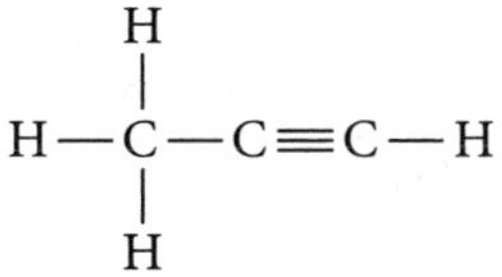

a.

H   H
| |
H—C—C—H
| |
H   H

_____

b.

H   H   H   H   H
| | | | |
H—C—C—C—C—C—H
| | | | |
H   H | H   H
        H—C—H
        |
        H

_____

c.

H
|
H—C—C≡C—H
|
H

_____

d.

        H
        |
        H—C—H
        |
H | H   H
| | | |
H—C———C———C—C—H
| | | |
H | H   H
        H—C—H
        |
        H

_____

The structure for benzene, C_6H_6, may be represented in two different ways, depending on the arrangement of the electrons:

Such structures are called *resonance* structures and are indicated by a double-headed arrow.

Often, either of these resonance structures is used to represent benzene. For simplicity, however, benzene is usually abbreviated as

280

What is the abbreviated structure for the following?

a.

a.

b.

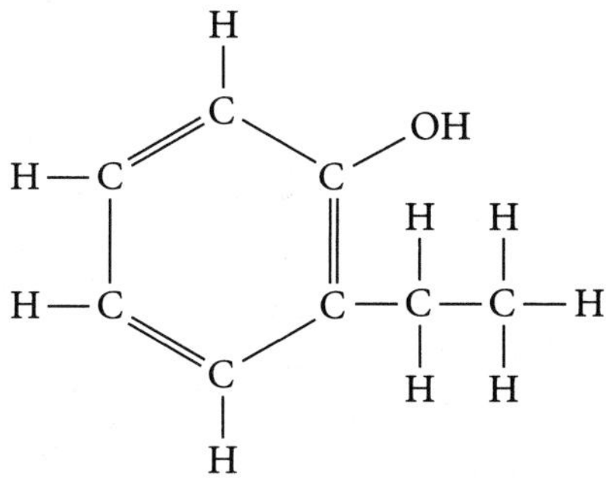

b.

281

Draw the complete structure for each of the following simplified formulas:

a. $CH_3CH_2CH_3$

a.

b. $CH_3CH = CHCH_3$

b.

c.

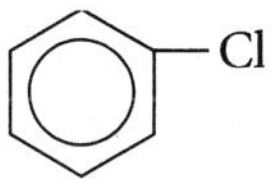

c.

or

d. $CH_3CH(CH_3)CH_2CH_2CH_3$

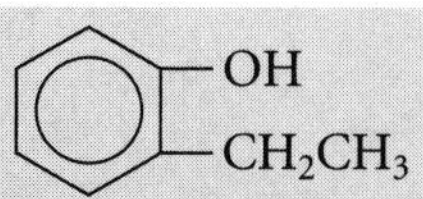

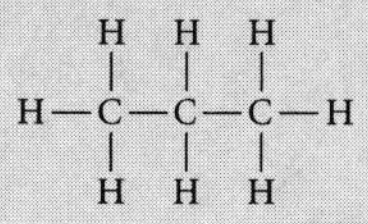

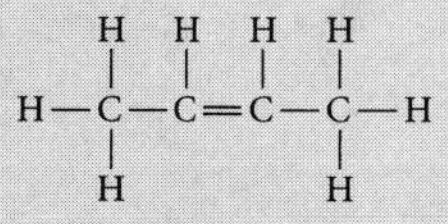

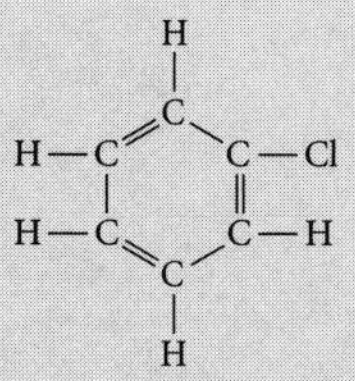

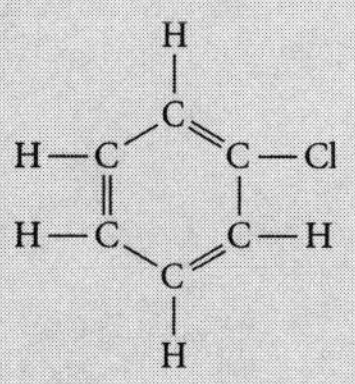

d.

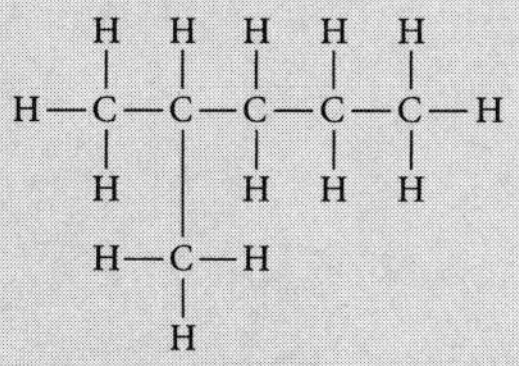

Polar and Nonpolar Covalent Bonds 9

282

Electronegativity is the attraction of an atom for electrons. The following table indicates the electronegativities of several elements.

F	4.0	Br	2.8	Al	1.5
O	3.5	C	2.5	Ca	1.0
Cl	3.1	H	2.1	Na	0.9
N	3.0				

Note that nonmetals have higher electronegativities than do metals. The difference in atoms' electronegativities determines the type of bond that can form between them.

When atoms of two nonmetals combine, the difference in electronegativities is small or zero and a covalent bond results.

When atoms of a metal and a nonmetal react, the difference in electronegativities is relatively larger and an ionic bond results.

a. Which of these elements has the greatest attraction for electrons?

———————

b. Which of these elements has the least attraction for electrons?

———————

a. F
b. Na

283

In the compound Cl — Cl, or Cl:Cl, there is a pair of shared electrons between the chlorine atoms. Is there a difference in electronegativity between the two chlorine atoms? ———————

no

a. Cl
b. closer to

284

Because the two chlorine atoms have the same electronegativity, the electrons should be shared equally between them. That is, the electrons should not be closer to one chlorine atom than to the other. Such a bond is called a *nonpolar covalent bond*, and such a compound is called a *nonpolar compound*. Nonpolar means that there is no negative end and no positive end to the bond or to the molecule.

a.   In the covalent compound HCl, H—Cl, or H:Cl, which element has the greater electronegativity? _____

b.   Because Cl is more electronegative than H, the shared electrons should be _____ (closer to/farther from) the chlorine.

a. yes
b. polar
c. polar

285

Because the shared electrons are closer to the Cl than to the H, the Cl will have a partial negative charge and the H will have a partial positive charge. This property may be indicated by a δ (Greek delta) sign.

$$\overset{\delta+\ \ \delta-}{HCl}$$

Note that it is a partial charge, not an ionic charge.

a.   Is one end of the molecule more negative than the other end?

_____

b.   The covalent bond between the H and the Cl is _____ (polar/nonpolar).

c.   Is the HCl molecule polar or nonpolar? _____

nonpolar

286

Consider the covalent compound XY. If both atoms have the same electronegativity, will the covalent bond between them be polar or nonpolar? _____

287

If there is a difference in electronegativity between X and Y, then the covalent bond between them will be _____ and the compound XY will be _____.

288

Now consider the covalent compound CO_2, or O::C::O

a. Which atom is more electronegative? _____ Which is less electronegative? _____

b. Which atom will have a partial + charge? _____ Which will have a partial − charge? _____

289

The CO_2 molecule may be represented as a linear molecule.

$$\overset{\delta-}{O}::\overset{\delta+}{C}::\overset{\delta-}{O}$$

Both ends of the CO_2 molecule have a partial _____ (positive/negative) charge.

290

a. Does the CO_2 molecule have a positive and a negative end? _____

b. Is CO_2 a polar or nonpolar compound? _____

nonpolar

291

In general, if a covalent compound is symmetrical, it will be _____ (polar/nonpolar).

H; O

292

In the H_2O molecule, which contains covalent bonds, which atom has a partial positive charge? _____

Which has a partial negative charge? _____

b; a

293

The structure of the water molecule may be represented as either

a. $\overset{\delta+}{H}—\overset{\delta-}{O}—\overset{\delta+}{H}$ or   b. $\overset{\delta+}{H}—\overset{\delta-}{O}$
$\diagdown$
$H^{\delta+}$

The structure in a is theoretical; it does not exist.

Which structure, a or b, represents a polar compound? _____

Which represents a nonpolar compound? _____

b

294

The water molecule has been found to be a polar molecule. Which of the structures in problem 293 more accurately depicts the water molecule?

_____

295

One general rule of solubility is that *like dissolves like.* That is, polar liquids dissolve polar compounds, and nonpolar liquids dissolve nonpolar compounds. Is water a polar or a nonpolar liquid? _____

polar

296

a.   Should NaCl, a polar (ionic) compound, dissolve in water? _____

b.   Should benzene, a nonpolar compound, dissolve in water? _____

c.   Should acetone, a nonpolar liquid, dissolve in benzene, a nonpolar liquid? _____

a. yes
b. no
c. yes

Functional Groups in Organic Compounds 10

Functional groups impart certain sets of properties to compounds. One example of such a functional group is — *OH group*.

297

Groups that impart certain sets of properties to organic compounds are called _____ groups.

THE HYDROXYL GROUP

298

The —OH group in an ionic compound is called a hydroxide ion.

$$NaOH \longrightarrow Na^+ + OH^-$$

Ionic

The —OH group in a covalent compound is called a *hydroxyl* group; it does not ionize.

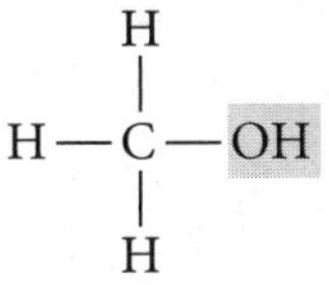

$$H - \overset{\displaystyle H}{\underset{\displaystyle H}{C}} - OH$$

Covalent, not ionized

Is a solution of NaOH an electrolyte? _____

functional

yes

no

299

Consider this compound:

$$
\begin{array}{c}
\text{H} \\
| \\
\text{H}-\text{C}-\text{OH} \\
| \\
\text{H}
\end{array}
$$

Is a solution of this compound an electrolyte? _____

The —OH group has the bonds —O—H. It is frequently written as —OH; the bond between the hydrogen and the oxygen is understood and is not written. Note that the oxygen atom still has two bonds attached to it.

In general, covalent compounds containing one or more —OH groups are called *alcohols*.

all of them

300

Which of the compounds a through e contains a hydroxyl group?

_____

a.

$$
\begin{array}{ccc}
\text{H} & \text{H} & \text{H} \\
| & | & | \\
\text{H}-\text{C}-\text{C}-\text{C}-\text{OH} \\
| & | & | \\
\text{H} & \text{H} & \text{H}
\end{array}
$$

b.

$$
\begin{array}{c}
\text{H} \\
| \\
\text{H}-\text{C}-\text{H} \\
\text{H} \quad | \quad \text{H} \\
| \quad | \quad | \\
\text{H}-\text{C}-\text{C}-\text{C}-\text{H} \\
| \quad | \quad | \\
\text{H} \quad \text{OH} \quad \text{H}
\end{array}
$$

c. $CH_3CH(OH)CH_3$

d. $CH_3CH_2CH_2CH(CH_3)CH_2OH$

e. ⬡—CH_2CH_2OH
        —CH_3

301

Glycerol, $C_3H_8O_3$, is one of the primary constituents of fats. Its structure is

$$
\begin{array}{ccc}
\text{H} & \text{H} & \text{H} \\
| & | & | \\
\text{H}-\text{C}-\text{C}-\text{C}-\text{H} \\
| & | & | \\
\text{OH} & \text{OH} & \text{OH}
\end{array}
$$

Does glycerol contain hydroxyl groups? _____

> yes

302

The names of alcohols are written with the ending *ol*. CH_3OH is methanol. CH_3CH_2OH is ethanol.

If the name of a compound ends in *ol*, that compound contains a _____ group and is called a(n) _____.

> hydroxyl; alcohol

303

Which of the following compounds is(are) an alcohol? _____

a.   cholesterol

b.   tocopherol (vitamin E)

c.   thiamine (vitamin B_1)

d.   cortisone

e.   retinol (vitamin A)

> a, b, e

304

Alcohols contain which functional group? _____

> hydroxyl

THE CARBONYL GROUP

A second functional group is the *carbonyl* group $C = O$.

yes

305

Look at the following structures:

$$\textbf{a. } CH_3CH_2CH_2 \overset{\overset{\displaystyle H}{|}}{C} = O \quad \text{and} \quad \textbf{b. } CH_3 \underset{\underset{\displaystyle O}{\|}}{C} CH_2CH_3$$

Do both contain a carbonyl group? _____

aldehyde; ketone

306

In compound a, the carbonyl group is at the end of the carbon chain; in compound b, the carbonyl group is not at the end of the carbon chain.

Compounds in which the carbonyl group is at the end of the carbon chain are called *aldehydes*.

Compounds in which the carbonyl group is not at the end of the carbon chain are called *ketones*.

Compound a is an example of a(n) _____ (aldehyde, ketone).

Compound b is an example of a(n) _____ (aldehyde, ketone).

carbonyl

307

Note that the functional group of an aldehyde may be written as

$$CH_3CH_2CH_2 \overset{\overset{\displaystyle H}{|}}{C} = O \quad \text{or} \quad CH_3CH_2CH_2CHO$$

and the functional group of a ketone may be written as

$$CH_3 \underset{\underset{\displaystyle O}{\|}}{C} CH_2CH_3 \quad \text{or} \quad CH_2COCH_2CH_3$$

The $C = O$ functional group is called the _____ group.

308

Both aldehydes and ketones contain which functional group?

_____

309

In an aldehyde, the carbonyl group is where on the carbon chain?

_____

310

In a ketone, where is the carbonyl group located? _____

311

Aldehydes have names ending in *al;* ketones have names ending in *one.*

Compound a in problem 305 is called *butanal.* The ending *al* indicates that butanal is what type of compound? _____

What functional group does it contain? _____

312

Compound b in problem 305 is called *butanone.*

Butanone is what type of compound? _____

What functional group does it contain? _____

carbonyl

at the end

not at the end of the chain

aldehyde; carbonyl

ketone; carbonyl

The following compounds contain which functional group(s)?

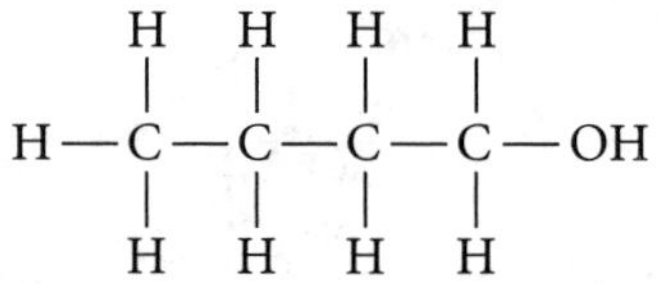

hydroxyl

313

H H H H
| | | |
H—C—C—C—C—OH
| | | |
H H H H

_____

carbonyl

314

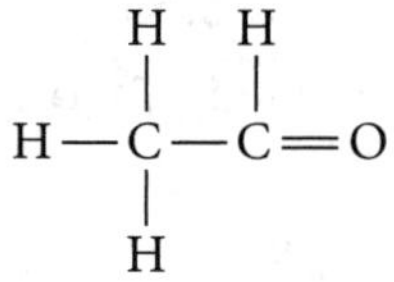

H H
| |
H—C—C=O
|
H

_____

carbonyl

315

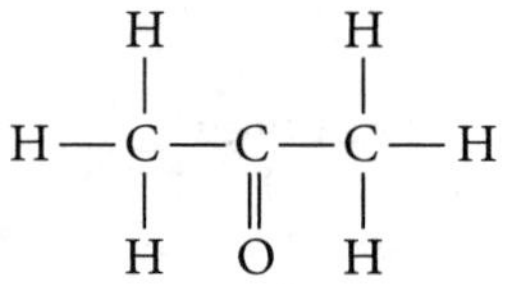

H   H
| |
H—C—C—C—H
| || |
H   O   H

_____

carbonyl

316

⬡—CHO

_____

317

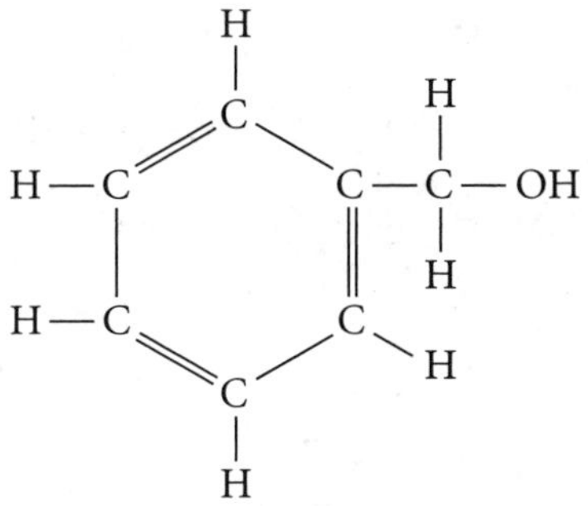

hydroxyl

_____

Identify the functional groups in the following compounds and identify the type of compound as alcohol, aldehyde, or ketone.

318

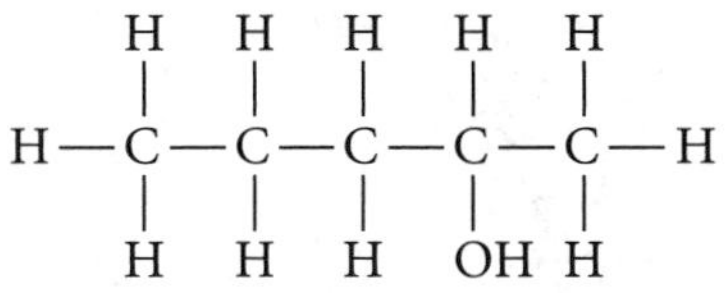

hydroxyl; alcohol

_____

319

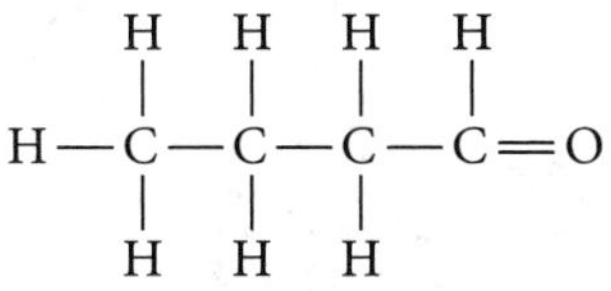

carbonyl; aldehyde

_____

carbonyl; ketone

320

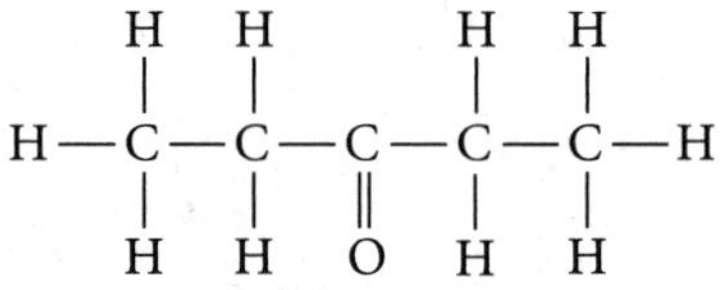

$$H-\underset{\underset{H}{|}}{\overset{\overset{H}{|}}{C}}-\underset{\underset{H}{|}}{\overset{\overset{H}{|}}{C}}-\underset{\underset{O}{\|}}{C}-\underset{\underset{H}{|}}{\overset{\overset{H}{|}}{C}}-\underset{\underset{H}{|}}{\overset{\overset{H}{|}}{C}}-H$$

_____

hydroxyl; alcohol

321

$CH_3CH(OH)CH_3$

_____

carbonyl; ketone

322

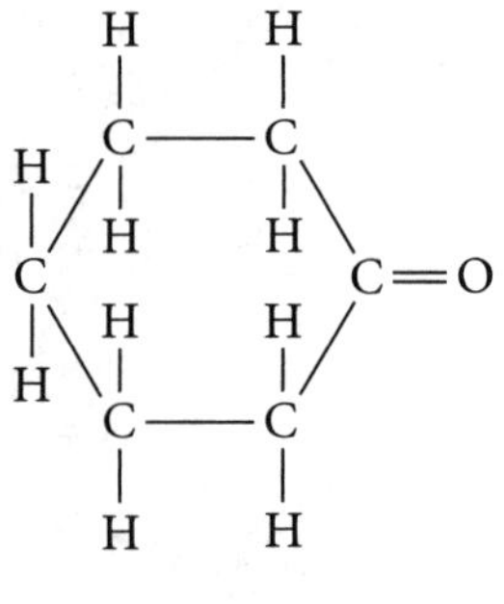

_____

carbonyl; ketone

323

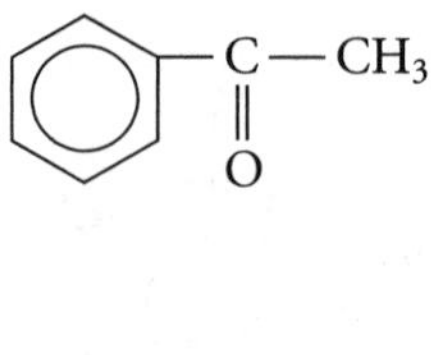

_____

324

$$CH_3 - \underset{\underset{O}{\|}}{C} - CH_2 - CH_2 - CHO$$

_____

carbonyls; aldehyde
and ketone

325

$$H - \underset{\underset{H}{|}}{\overset{\overset{H}{|}}{C}} - \underset{\underset{H}{|}}{\overset{\overset{H}{|}}{C}} - \underset{\underset{OH}{|}}{\overset{\overset{H}{|}}{C}} - \underset{\underset{H}{|}}{\overset{\overset{H}{|}}{C}} - C = O$$

_____

hydroxyl; carbonyl
alcohol and aldehyde

326

CHO
C — CH₃
O

_____

carbonyls; aldehyde
and ketone

327

The names of most ketones end in *one*. An example of a ketone is
acetone. Which of the following substances are ketones? _____
Alcohols? _____ Aldehydes? _____

a. corticosterone   b. pyridoxal (a B-vitamin)
c. cortisol   d. androsterone
e. estriol   f. pentanal
g. pregnandiol   h. retinal

ketones: a, d
alcohols: c, e, g
aldehydes: b, f, h

THE CARBOXYL GROUP

acid

328

A third functional group is the *carboxyl* group, —COOH or

$$\begin{array}{c} O \\ \parallel \\ -C-OH \end{array}$$

The carboxyl group ionizes as follows:

$$\begin{array}{c} O \\ \parallel \\ -C-OH \end{array} \longrightarrow \begin{array}{c} O \\ \parallel \\ -C-O^- \end{array} + H^+$$

Because the carboxyl group yields H^+ ions, it is a(n) _____.

Because organic acids contain a carboxyl group, they are also called *carboxylic acids.*

a. an aldehyde; carbonyl
b. a carboxylic acid;
carboxyl

329

The —COOH group may also be written as

$$\begin{array}{ccc} O & & OH \\ \parallel & & \mid \\ -C-OH & or & -C=O \end{array}$$

a. A compound with a —CHO group is called a(n) _____ and contains which functional group? _____

b. A compound with a —COOH group is called a(n) _____ and contains which functional group? _____

330

Citric acid, α-ketoglutaric acid, succinic acid, and oxaloacetic acid are a part of the Krebs cycle, the human body's primary metabolic cycle. What functional group should they all have in common? _____

carboxyl, or —COOH

331

What functional groups are present in the following compounds?

a. pyruvic acid, produced during the oxidation of glucose _____

$$CH_3 - \underset{\underset{O}{\|}}{C} - COOH$$

b. lactic acid, formed during the fermentation of milk _____

$$CH_3 - \underset{\underset{OH}{|}}{CH} - COOH$$

c. tartaric acid, found in grapes _____

$$\begin{array}{c} OH \\ | \\ H - C - COOH \\ | \\ H - C - COOH \\ | \\ OH \end{array}$$

a. carbonyl and carboxyl (organic acid)
b. hydroxyl and carboxyl
c. hydroxyl and carboxyl

THE AMINO GROUP

Another functional group is the amino group —NH_2

Compounds containing an amino group are called *amines*.

yes; yes

| 332 |

Consider the following compound:

$$CH_3NH_2$$

Methylamine

Does it contain an amino group? _____

Is it an amine? _____

amino and carboxyl

| 333 |

Consider the compound glycine:

$$\begin{array}{c} COOH \\ | \\ CH_2NH_2 \end{array}$$

What functional group(s) does it contain? _____

yes, yes

| 334 |

Is glycine an amine? An acid? _____

Compounds that contain both an amino group and a carboxyl group are called *amino acids*.

335

Which of the following compounds contain(s) an amino group?

_____

a. CH_3NH_2

b.

—$CH_2CH_2NH_2$

c. $CH_3CHCH_2CH_3$
       |
       NH_2

d. CH_2—CH—CH_2—CH_3
       | |
       NH_2 NH_2

336

a.   An organic acid is a compound containing a(n) _____ group.

b.   A compound containing an —NH_2 group is called _____.

c.   A compound containing an —OH group is called _____.

d.   A compound containing a —CHO group is called _____.

e.   A compound containing a C—C—C group is called _____.
                                      ‖
                                      O

a. —COOH (carboxyl)
b. an amine
c. an alcohol
d. an aldehyde
e. a ketone

Identify the functional groups in the following problems.

amino

337

$$H-\underset{\underset{H}{|}}{\overset{\overset{H}{|}}{C}}-\underset{\underset{NH_2}{|}}{\overset{\overset{H}{|}}{C}}-\underset{\underset{H}{|}}{\overset{\overset{H}{|}}{C}}-H$$

hydroxyl and carbonyl

338

$$H-\underset{\underset{H}{|}}{\overset{\overset{H}{|}}{C}}-\underset{\underset{OH}{|}}{\overset{\overset{H}{|}}{C}}-\underset{\underset{H}{|}}{\overset{\overset{H}{|}}{C}}-\overset{\overset{H}{|}}{C}=O$$

hydroxyl and carbonyl

339

$$H-\underset{\underset{H}{|}}{\overset{\overset{H}{|}}{C}}-\underset{\underset{OH}{|}}{\overset{\overset{H}{|}}{C}}-\underset{\underset{H}{|}}{\overset{\overset{H}{|}}{C}}-\underset{\underset{O}{||}}{C}-\underset{\underset{H}{|}}{\overset{\overset{H}{|}}{C}}-H$$

amino and hydroxyl

340

$$H-\underset{\underset{H}{|}}{\overset{\overset{H}{|}}{C}}-\underset{\underset{H}{|}}{\overset{\overset{H}{|}}{C}}-\underset{\underset{NH_2}{|}}{\overset{\overset{H}{|}}{C}}-\underset{\underset{H}{|}}{\overset{\overset{H}{|}}{C}}-\underset{\underset{H}{|}}{\overset{\overset{H}{|}}{C}}-OH$$

341

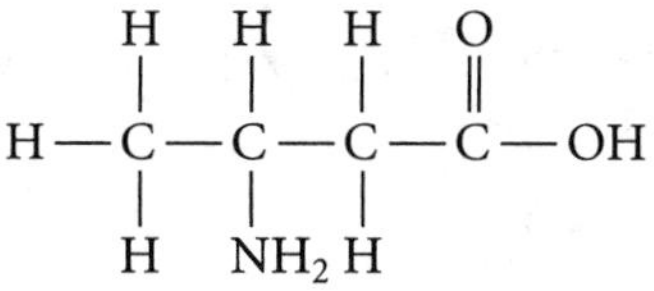

$$
\begin{array}{c}
\;\;\;\;\;\;\;\;\;\; H \;\;\;\; H \;\;\;\; H \;\;\;\; O \\
\;\;\;\;\;\;\;\;\;\; | \;\;\;\;\;\;\; | \;\;\;\;\;\;\; | \;\;\;\;\;\;\; \| \\
H - C - C - C - C - OH \\
\;\;\;\;\;\;\;\;\;\; | \;\;\;\;\;\;\; | \;\;\;\;\;\;\; | \\
\;\;\;\;\;\;\;\;\;\; H \;\;\; NH_2 \;\; H
\end{array}
$$

amino and carboxyl

342

COOH

OH

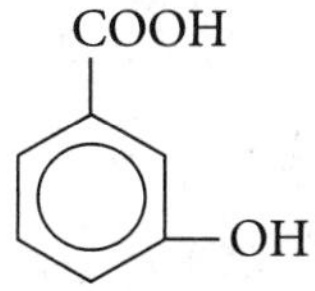

carboxyl and hydroxyl

343

$CH_3CH(NH_2)COOH$

amino and carboxyl
(amino acid)

344

$CH_3CCH_2CH_2NH_2$
$\underset{O}{\overset{\|}{}}$

carbonyl and amino

345

$CH_3CH_2CH(OH)COOH$

hydroxyl and carboxyl

THE SULFHYDRYL GROUP

The —SH group is called the sulfhydryl group. A compound containing a sulfhydryl group is called a *thiol*.

346

sulfur and hydrogen

The sulfhydryl group contains which elements?

347

yes

Consider the following compound:

$$CH_3CH_2SH$$

Ethanethiol

Does it contain a sulfhydryl group? _____

348

carboxyl, amino, and sulfhydryl

Cysteine, a non-essential amino acid has the following structure:

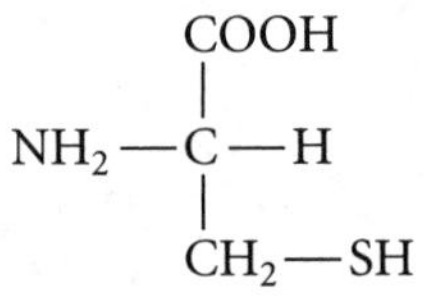

$$
\begin{array}{c}
COOH \\
| \\
NH_2-C-H \\
| \\
CH_2-SH
\end{array}
$$

Cystine

Which functional groups does it contain? _____

DISULFIDE BRIDGES

Disulfide bridges contain a —S—S— group.

349

What does the term disulfide refer to?

The following compound, cystine, contains a disulfide bridge.

$$
\begin{array}{ccc}
\text{COOH} & & \text{COOH} \\
| & & | \\
\text{NH}_2\!-\!\text{CH} & \text{NH}_2\!-\!\text{CH} \\
| & & | \\
\text{CH}_2\!-\!\text{S}\!-\!\text{S}\!-\!\text{CH}_2 &
\end{array}
$$

Cystine

350

The prefix *di* indicates the number _____. Sulfide refers to the element _____. So, a disulfide bridge contains _____ (how many?) atoms of the element _____.

351

Note the following abbreviated structure of human insulin.

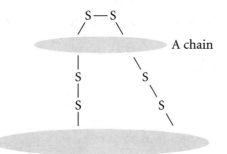

What type of bonds or linkages holds the A and B chains together?

_____

How many bonds of this type are present in human insulin? _____

two sulfides

2; sulfur; 2; sulfur

disulfide; 3

THE PHOSPHATE GROUP

The —OPO_3H_2 group is called a *phosphate* group. This group may also be written as

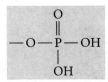

352

Compounds containing a phosphate group are called phosphates. Is the following compound a phosphate?

$$\begin{array}{ccccc}
 & H & O & & O \\
 & | & \| & & \| \\
H- & C- & C- & O- & P-OH \\
 & | & & & | \\
 & H & & & OH
\end{array}$$

353

In ATP, adenosine triphosphate, the P indicates which functional group?

_____

354

The prefix *tri* indicates how many phosphate groups as being present in ATP? _____

yes

phosphate

3

355

Look at the following ladder representation of DNA. The S along the backbone of the chain indicates a sugar. The P indicates which functional group? _____

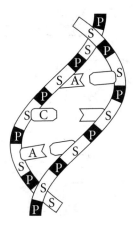

phosphate

356

The following structure represents adenosine monophosphate (AMP). Draw a circle around the phosphate group. This compound also contains an amino and two hydroxyl groups. Draw a square around the amino group, and a rectangle around the hydroxyl groups.

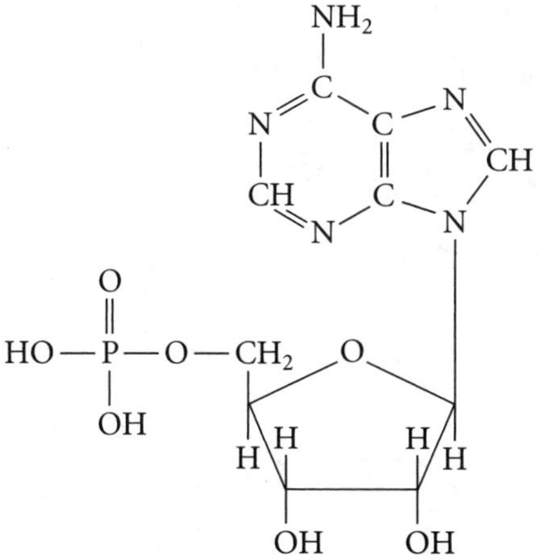

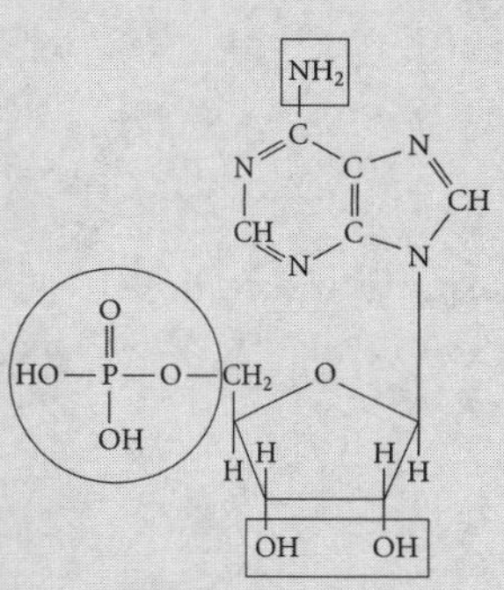

SUMMARY OF
FUNCTIONAL GROUPS

Name the functional groups present at the following compounds.

carbonyl and hydroxyls

357

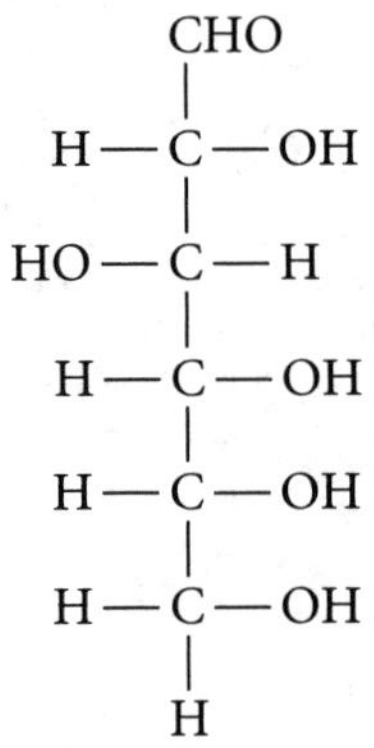

$$
\begin{array}{c}
\mathrm{CHO} \\
| \\
\mathrm{H-C-OH} \\
| \\
\mathrm{HO-C-H} \\
| \\
\mathrm{H-C-OH} \\
| \\
\mathrm{H-C-OH} \\
| \\
\mathrm{H-C-OH} \\
| \\
\mathrm{H}
\end{array}
$$

Glucose

**amino and carboxyl
(amino acid)**

358

$$CH_3CH(NH_2)COOH$$

Alanine

359

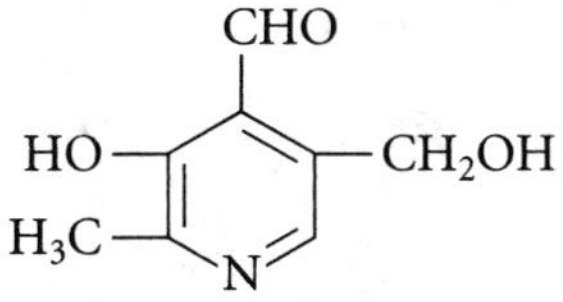

CHO

HO

H$_3$C

CH$_2$OH

N

Pyridoxal
(one of the B-vitamins)

carbonyl and hydroxyl

360

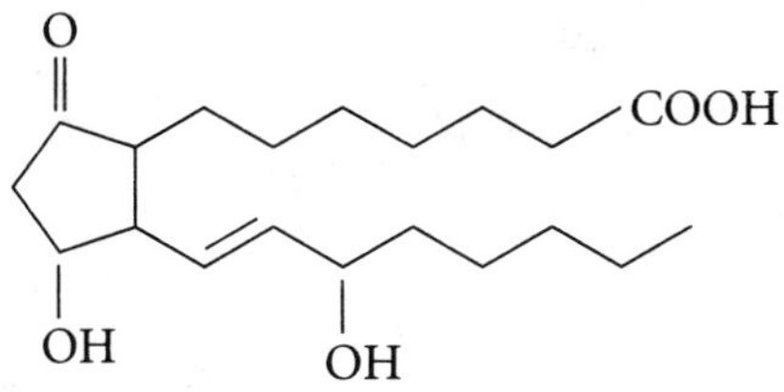

O

COOH

OH

OH

Prostaglandin

two hydroxyl groups, a
carboxyl group, and a
carbonyl group

361

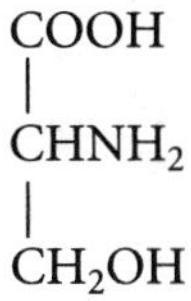

COOH
|
CHNH$_2$
|
CH$_2$OH

Serine

a hydroxyl group, an
amino group, and a
carboxyl group

carbonyl, hydroxyl, and phosphate groups

362

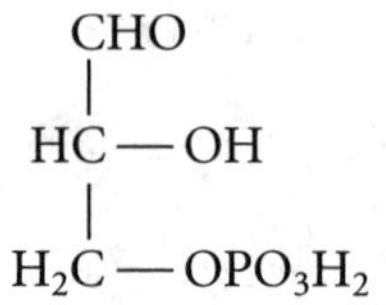

CHO
|
HC — OH
|
H_2C — OPO_3H_2

Glyceraldehyde-3-P

---

carboxyl, amino

363

COOH

NH_2

---

a sulfhydryl group

364

$CH_3CH_2CH_2SH$

---

two carboxyl groups, two amino groups, and one disulfide bridge

365

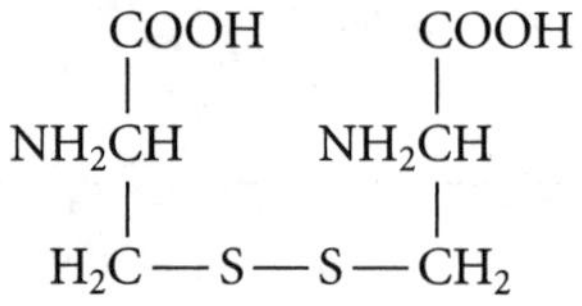

COOH   COOH
| |
NH_2CH NH_2CH
| |
H_2C — S — S — CH_2

| 366 | a carboxyl group, an amino group, and a sulfhydryl group |

$$COOH$$
$$|$$
$$NH_2CH$$
$$|$$
$$CH_2 — SH$$

| 367 | disulfide bridge |

The following diagram illustrates part of the shape of a protein molecule.

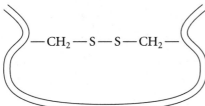

What type of bond or bridge determines the conformation of such a protein? _____

What type of functional group(s) is(are) present in each of the following compounds?

| 368 | carbonyl and carboxyl |

*alpha*ketoglutaric acid _____

| 369 | sulfhydryl |

methanethiol _____

amino and carboxyl

370

essential aminoacids _____

carbonyl

371

acetone _____

carboxyl

372

*trans*fatty acids _____

amino and carboxyl

373

*para*aminobenzoic acid _____

hydroxyl

374

pregnanediol _____

carbonyl

375

testosterone _____

Hydrogen Bonds $\quad$ **11**

376

Consider two water molecules whose structures and partial charges are

$$\overset{\delta+}{H}-\overset{\delta-}{O} \qquad \text{and} \qquad \overset{\delta+}{H}-\overset{\delta-}{O}$$
$$\backslash_{H^{\delta+}} \qquad\qquad\qquad \backslash_{H^{\delta+}}$$

Which partial charges should attract each other? _____

the negative charge of one water molecule and the positive charge of the other

weak

377

Water molecules can form a weak bond between the H of one molecule and the O of another. Such a bond, called a *hydrogen bond*, is caused by the attraction of the partially positive H of one water molecule and the partially negatively charged O (more electronegative) of another water molecule. Hydrogen bonds may be indicated by dotted lines:

Is a hydrogen bond a strong or a weak bond? _____

hydrogen bond

378

If many water molecules are bonded together in a similar manner, what type of bond should hold them together? _____

379

Hydrogen bonds can also occur between the H attached to an amino group and an O that is part of a carboxyl group. Such hydrogen bonds are present in proteins and in DNA.

Note the following diagram, which illustrates a very small part of a protein molecule. (R represents a carbon chain.)

hydrogen bond

$$
\begin{array}{c}
\diagup \text{C}=\text{O}\cdots\text{H}-\text{N} \diagup \qquad \diagdown \text{C}=\text{O}\cdots\text{H}-\text{N} \diagup \\
\text{H}-\text{N} \qquad \text{C}=\text{O}\cdots\text{H}-\text{N} \qquad \text{C}= \\
\text{CHR} \qquad \text{CHR} \qquad \text{CHR} \qquad \text{CHR} \\
\text{O}=\text{C} \qquad \text{N}-\text{H}\cdots\text{O}=\text{C} \qquad \text{N}- \\
\text{N}-\text{H}\cdots\text{O}=\text{C} \qquad \text{N}-\text{H}\cdots\text{O}=\text{C} \\
\text{CHR} \qquad \text{CHR} \qquad \text{CHR} \qquad \text{CHR} \\
\text{C}=\text{O}\cdots\text{H}-\text{N} \qquad \text{C}=\text{O}\cdots\text{H}-\text{N} \\
\text{H}-\text{N} \qquad \text{C}=\text{O}\cdots\text{H}-\text{N} \qquad \text{C}= \\
\text{CHR} \qquad \text{CHR} \qquad \text{CHR} \qquad \text{CHR} \\
\text{O}=\text{C} \qquad \text{N}-\text{H}\cdots\text{O}=\text{C} \qquad \text{N}- \\
\text{N}-\text{H}\cdots\text{O}=\text{C} \qquad \text{N}-\text{H}\cdots\text{O}=\text{C} \\
\text{CHR} \qquad \text{CHR} \qquad \text{CHR} \qquad \text{CHR} \\
\text{C}=\text{O}\cdots\text{H}-\text{N} \qquad \text{C}=\text{O}\cdots\text{H}-\text{N} \\
\text{H}-\text{N} \qquad \text{C}=\text{O}\cdots\text{H}-\text{N} \qquad \text{C}=
\end{array}
$$

Pleated sheet structure of a protein

What type of bond holds the sections together? _____

In this type of structure the C = O is derived from a *carboxyl* group and the N — H from an *amino* group.

380

amino; carboxyl

The hydrogen bonds in the diagram in problem 379 exist between the H attached to a(n) _____ group and the oxygen attached to a(n) _____ group.

hydrogen bond

381

Hydrogen bonding accounts for the pleated sheet structure of some proteins.

Some proteins have a helical structure, as shown in the following diagram.

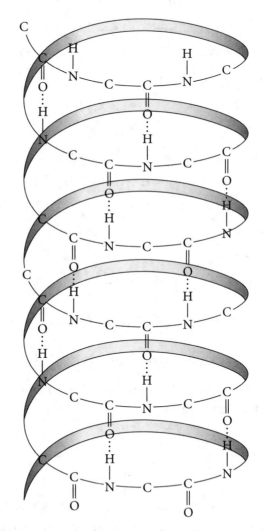

Helical structure of a protein

What type of bond holds the coil in its designated shape? _____

382

In both the pleated sheet and helical structures of proteins, the hydrogen bonds present exist between the H attached to a(n) _____ and the O that is part of a(n) _____ group.

383

Consider the double helical shape of DNA in the following diagram.

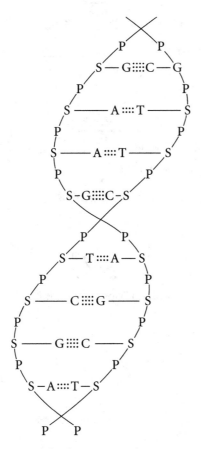

Helical structure of DNA

What type of bond holds the two halves together? _____

2; hydrogen bonds

384

The DNA structure contains the following substances: adenine (A), guanine (G), cytosine (C), and thymine (T).

How many bonds are present between A and T? _____

What type of bonds are these? _____

3; hydrogen bonds

385

How many bonds are present between C and G? _____

What type of bonds are these? _____

weak

386

When the DNA molecule replicates, the two halves come apart, breaking the bonds holding them together.

Are hydrogen bonds strong or weak bonds? _____

easy

387

Should it be easy or difficult to "open up" the DNA chain? _____

Isomers are compounds that have the same atoms (same molecular formulas) but different three-dimensional arrangement of those atoms. Hence, isomers have different properties from one another.

Isomers can be classified as

a. structural isomers

b. geometric isomers

c. enantiomers

STRUCTURAL ISOMERS

388

Consider the following two compounds.

$$CH_3CH_2CH_2CH_3 \quad \text{and} \quad \overset{\displaystyle CH_3}{\underset{}{\overset{|}{CH_3CHCH_3}}}$$

a   b

What is the molecular formula for compound a? _____

What is the molecular formula for compound b? _____

$C_4H_{10}; C_4H_{10}$

no

C₂H₆O; C₂H₆O; yes; no

389

Compounds a and b are structural isomers because they have the same molecular formula but different three-dimensional structures.

Would compounds a and b have identical properties? _____

390

Consider the following two structures.

$$CH_3OCH_3 \qquad CH_3CH_2OH$$

Methyl ether   Ethanol

What is the molecular formula for methyl ether? _____

What is the molecular formula for ethanol? _____

Are methyl ether and ethanol structural isomers? _____

Do methyl ether and ethanol have the same properties? _____

391

Are the following pairs structural isomers?

a. _____

$$CH_3CH_2CCH_2CH_3 \quad \text{and} \quad CH_3CH_2CH_2CH_2CH$$
with the first having a $\parallel$ O below the third carbon, and the second having O double-bonded at the end

b. _____

$$CH_2CH_2CH_2 \quad \quad CH_3CHCH_2$$
$$\quad | \quad \quad | \quad \quad \quad \quad | \quad \; |$$
$$\quad OH \quad \; OH \quad \quad \quad OH\,OH$$

c. _____

$$CH_3CHCH_2CH_2COOH \quad \text{and} \quad CH_3CH_2CH_2CHCOOH$$
$$\quad \; | \quad \quad \quad \quad \quad \quad \quad \quad \quad \quad \quad | $$
$$\quad NH_2 \quad \quad \quad \quad \quad \quad \quad \quad \quad \quad NH_2$$

d. _____

COOH and COOH

benzene ring with —OH   and   benzene ring with —NH$_2$

a. yes
b. yes
c. yes
d. no

GEOMETRIC ISOMERS

Geometric isomers have the same molecular formula but different three-dimensional structures due to the presence of a double bond. A double bond is inflexible and hence does not allow the atoms to rotate freely. Such a restriction in rotation can account for a difference in biologic activity for those isomers.

a; b

$$\boxed{392}$$

Consider the following arrangement of carbon atoms

$$\begin{array}{cc} X & Y \\ | & | \\ -C & = C - -- \end{array}$$

where **X** and **Y** are the same or different atoms or groups of atoms attached to the double bond in a carbon chain.

Two possible arrangements of **X** and **Y** in such a compound are

a.
$$\begin{array}{cc} X & Y \\ | & | \\ --- -C & = C - \end{array}$$

b.
$$\begin{array}{cc} X & \\ | & \\ -C = C - \\ & | \\ & Y \end{array}$$

Note that in such geometric isomers, the **X** and **Y** groups can be on the "same side" of the double bond or on "opposite sides" of the double bond.

If both **X** and **Y** groups are on the same side of the double bond, then the geometric isomer is called a *cis* isomer. If the **X** and **Y** groups are on opposite sides of the double bond, then such a compound is called a *trans* isomer.

Which of the above is an example of a *cis* isomer? _____

Which of the above is an example of a *trans* isomer? _____

Geometric isomers are also called cis-trans isomers.

393

Which of the following compounds is(are) a *cis* isomer? _____

Which is(are) a *trans* isomer? _____

a.

$$CH_3 - \overset{\overset{\displaystyle Cl}{|}}{C} = \overset{\overset{\displaystyle Cl}{|}}{C} - CH_2 - CH_3$$

b.

$$CH_3CH_2 - \underset{\underset{\displaystyle Cl}{|}}{C} = \overset{\overset{\displaystyle Cl}{|}}{C} - CH_2 - CH_2COOH$$

c.

$$CH_3 - \underset{\underset{\displaystyle CH_2CH_3}{|}}{C} = \overset{\overset{\displaystyle CH_2CH_3}{|}}{C} - CH_2 - CH_3$$

394

Suppose that a carbon compound contains two double bonds such as

$$C - C = C - C - C = C - C$$
$$\quad\;\;\textbf{a.}\qquad\qquad\textbf{b.}$$

If the double bond labeled **a** has substances attached on the same side of that double bond, would the compound be a *cis* or *trans* isomer?

_____

395

If the double bond labeled **b** also had a *cis* arrangement, then the isomer would be labeled a *cis*-cis isomer.

What type of structure would a *trans-trans* isomer be? _____

Would a pair of cis-trans isomers have the same properties?

_____

a; b and c

cis

one where each double bond would have a *trans* arrangement; no

ENANTIOMERS

Enantiomers are compounds with the same molecular formula but different three-dimensional structures that are nonsuperimposible mirror images of each other.

no

396

The mirror image of an object is the reflection of that object in a plane mirror.

If you hold your left hand in front of a plane mirror, you will see its mirror image, one that is identical to your right hand.

Left hand   Mirror
                                   image of
                                   left hand

Is your left hand superimposible on your right hand? _____

397

Look at the structure for glyceraldehyde and its mirror image.

$$
\begin{array}{ccc}
\text{CHO} & & \text{CHO} \\
| & & | \\
\text{H}-\text{C}-\text{OH} & & \text{HO}-\text{C}-\text{H} \\
| & & | \\
\text{CH}_2\text{OH} & & \text{CH}_2\text{OH}
\end{array}
$$

Glyceraldehyde   Mirror image of
                        Glyceraldehyde

These same two structures can be represented as

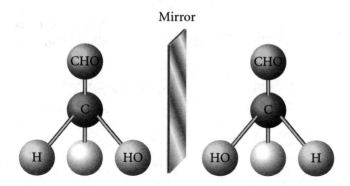

Mirror

Are these two structures superimposible? _____

398

Are these two compounds enantiomers? _____

399

Look at the following structure for glyceraldehyde.

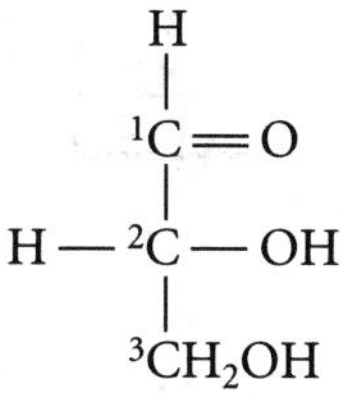

$$
\begin{array}{c}
H \\
| \\
{}^{1}C{=}O \\
| \\
H{-}{}^{2}C{-}OH \\
| \\
{}^{3}CH_2OH
\end{array}
$$

Carbon 2 has four different atoms or groups of atoms attached to it.
They are: _____, _____,
_____, and _____ (in any order)

no; no

400

Does carbon 1 have four different atoms or groups attached to it?

_____

Does carbon 3 have four different groups attached to it? _____

No, it does not have 4 different atoms attached to it.

401

A carbon with four different atoms or group of atoms attached to it is called an *asymmetric* carbon atom. Is carbon 1 in question 399 asymmetric? _____. Why or why not? _____

Enantiomers (also called optical isomers) occur because of the presence of one or more asymmetric carbon atoms.

contains; yes

402

Glyceraldehyde _____ (contains/does not contain) an asymmetric carbon. Should glyceraldehyde have enantiomers? _____.

no; no

403

Methane, CH_4, has the structure

$$H - \overset{\displaystyle H}{\underset{\displaystyle H}{\overset{|}{\underset{|}{C}}}} - H$$

Does methane have an asymmetric carbon? _____

Should methane have enantiomers? _____

404

The number of enantiomers (isomers) possible for a compound containing an asymmetric carbon atom is given by the formula

$$2^n$$

where n is the number of asymmetric carbon atoms present.

Glyceraldehyde has 1 asymmetric carbon atom. How many isomers are possible for glyceraldehyde? _____

405

Look at the following structure.

$$
\begin{array}{c}
^1CHO \\
| \\
H-^2C-OH \\
| \\
H-^3C-OH \\
| \\
^4CH_2OH
\end{array}
$$

Carbon 1 has which groups attached to it? _____

406

Carbon 2 has which groups attached to it? _____

407

Carbon 3 has which groups attached to it? _____

2^1 or 2

$$
-H, =O, \quad
\begin{array}{c}
H-C-OH \\
| \\
H-C-OH \\
| \\
CH_2OH
\end{array}
$$

$$
CHO, H, OH, \quad
\begin{array}{c}
H-C-OH \\
| \\
CH_2OH
\end{array}
$$

$$
\begin{array}{c}
CHO \\
| \\
H-C-OH, H, OH, CH_2OH
\end{array}
$$

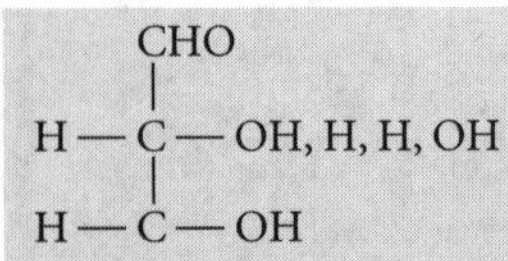

$$\begin{array}{c} \text{CHO} \\ | \\ \text{H}-\text{C}-\text{OH, H, H, OH} \\ | \\ \text{H}-\text{C}-\text{OH} \end{array}$$

2 and 3

2^2 or 4

4

408

Carbon 4 has which groups attached to it? _____

409

Which of the above indicated carbons (1, 2, 3, or 4) is(are) asymmetric? _____

410

The compounds indicated in problem 408 have how many optical isomers? _____

411

Glucose, $C_6H_{12}O_6$, has the following structure.

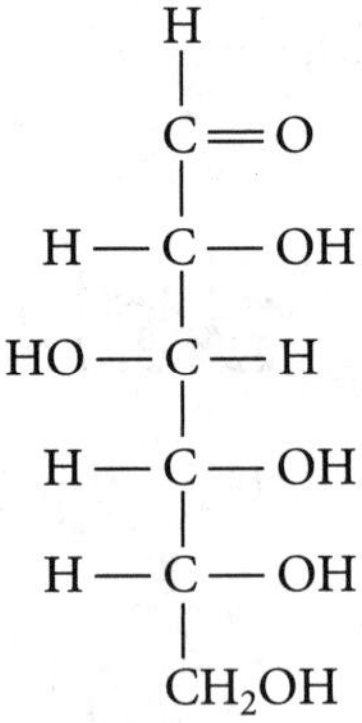

$$\begin{array}{c} \text{H} \\ | \\ \text{C}=\text{O} \\ | \\ \text{H}-\text{C}-\text{OH} \\ | \\ \text{HO}-\text{C}-\text{H} \\ | \\ \text{H}-\text{C}-\text{OH} \\ | \\ \text{H}-\text{C}-\text{OH} \\ | \\ \text{CH}_2\text{OH} \end{array}$$

Glucose has how many asymmetric carbons? _____

412

Glucose has how many isomers? _____

413

Glucose contains which functional group(s)? _____

2^4 or 16

carbonyl, hydroxyl

PART **III**

Biochemistry

Carbohydrates 13

Carbohydrates contain the elements carbon, hydrogen, and oxygen, and only those elements.* With few exceptions, a compound containing elements other than these three is not a carbohydrate.

The word *hydrate* refers to water. Many carbohydrates contain hydrogen and oxygen in the same ratio as in water, 2:1.

414

Which of the following compounds are carbohydrates?

___ C_6H_6 ___ $C_{12}H_{22}O_{11}$ ___ $C_5H_{10}O_5$
___ $C_6H_{12}O_6NS$ ___ C_2H_6O

$C_{12}H_{22}O_{11}$ and $C_5H_{10}O_5$

415

a. In the list of compounds in problem 414, why is $C_6H_{12}O_6NS$ not a carbohydrate? _____
_____

b. Why are C_6H_6 and C_2H_6O not carbohydrates? _____
_____
_____

a. $C_6H_{12}O_6NS$ contains nitrogen and sulfur, and carbohydrates contain only C, H, and O

b. C_6H_6 contains no oxygen, and C_2H_6O has the wrong ratio of H to O

---

* An exception to this generalization is *chitin*, a structural polysaccharide (carbohydrate) used by arthropods to build their exoskeletons and also used by some fungi. Chitin is composed of many glucose molecules $(C_6H_{12}O_6)$ modified with a nitrogen-containing compound.

a. one
b. one

all of them

5

416

Carbohydrates are divided into three types. One type of carbohydrate is the *monosaccharide*.

a.   The word *saccharide* means *simple sugar*. The prefix *mono* means

_____.

b.   Therefore, monosaccharides are simple sugars, each containing only _____ simple sugar.

417

Monosaccharides and disaccharides have names ending in *ose*. Which of the following are monosaccharides or disaccharides?

__ glucose   __ fructose   __ galactose
__ arabinose   __ erythrose

418

Monosaccharides are named according to the number of carbon atoms they contain. A pentose contains _____ carbon atoms.

An example of a pentose is ribose, $C_5H_{10}O_5$. The structure of ribose is

In this type of structure, the ring, consisting of carbon atoms at each corner and an oxygen atom where indicated, is considered to be in a plane perpendicular to the paper. The heavy line in the ring indicates the section closest to you; the light lines, the part farther from you. The — H and — OH (hydroxyl) groups are above and below the plane as indicated.

Ribose occurs in ribonucleic acid, RNA, which will be discussed later.

419

a. one less oxygen
b. one less oxygen

Compare the structures of ribose, $C_5H_{10}O_5$, and deoxyribose, $C_5H_{10}O_4$.

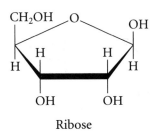
Ribose

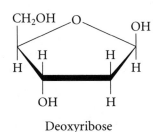
Deoxyribose

a. The prefix *deoxy* means _____.

b. Deoxyribose contains _____ than does ribose.

420

DNA

Ribose occurs in ribonucleic acid, RNA. Deoxyribose occurs in deoxyribonucleic acid, _____.

a. 5
b. 6

ose

421

a. A pentose is a monosaccharide containing _____ carbon atoms.

b. A hexose is a monosaccharide containing _____ carbon atoms.

A primary hexose in the body is glucose, $C_6H_{12}O_6$. The oxidation of glucose supplies much of the energy the body needs. The structure of glucose is

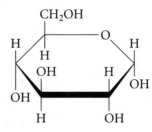

422

The second type of carbohydrate is the *disaccharide.* Both monosaccharides and disaccharides have names ending in _____.

423

The prefix *di* means *two*. Disaccharides are formed by the combination of two monosaccharides, with water also being produced. An enzyme is required for this reaction.

monosaccharide + monosaccharide $\xrightarrow{\text{enzyme}}$ disaccharide + water

Conversely, disaccharides react with water to produce two mono-saccharides.

disaccharide + water $\xrightarrow{\text{enzyme}}$ monosaccharide + _____

These reactions may be written as shown on the following page.

The bond holding the two halves of the disaccharide together is called a *glycosidic bond*, or a *glycosidic linkage*.

monosaccharide

424

When a disaccharide reacts with water, two monosaccharides are produced. Such a reaction is called *hydrolysis*, which is defined as the breaking apart of a molecule by reaction with water.

Digestion involves hydrolysis. During digestion, molecules are:

__ broken down into smaller molecules
__ built up into larger molecules

broken down into
smaller molecules

monosaccharides

425

Disaccharides that have undergone hydrolysis yield two _____.

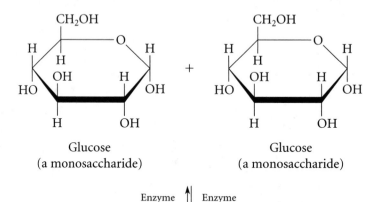

Glucose
(a monosaccharide)

+

Glucose
(a monosaccharide)

Enzyme   Enzyme
(hydrolysis)   (synthesis)

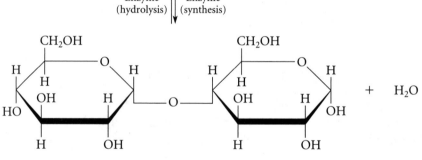

Maltose
(a disaccharide)

+ H_2O

Water

one; two

426

Monosaccharides are made up of _____ simple sugar(s).
Disaccharides are made up of _____ simple sugar(s).

427

The hydrolysis of a disaccharide may be written as:

$$C_{12}H_{22}O_{11} + H_2O \xrightarrow{\text{enzyme}} C_6H_{12}O_6 + C_6H_{12}O_6$$

Disaccharide + water   Monosaccharide + monosaccharide

Sucrose, $C_{12}H_{22}O_{11}$, is a disaccharide. The hydrolysis of sucrose yields the two _____, glucose and fructose.

428

There are three common disaccharides: *sucrose*, *maltose*, and *lactose*. These disaccharides all have the same molecular formula, $C_{12}H_{22}O_{11}$. Lactose, also called milk sugar, is found in milk. Maltose, or malt sugar, is found in sprouting grain. Sucrose, or cane sugar, is found in _____.

429

The third type of carbohydrate is the *polysaccharide*. The prefix *poly* means *many*, so polysaccharides, on hydrolysis, yield many _____.

430

Recall that hydrolysis means reaction with _____ to produce _____ (smaller/larger) molecules.

monosaccharides

sugarcane

monosaccharides

water; smaller

$C_{12}H_{22}O_{11}$

431

Starch is a polysaccharide. Its molecular formula is $(C_6H_{10}O_5)_n$, where n is some large number. Upon complete hydrolysis, starch yields many monosaccharides:

$$(C_6H_{10}O_5)_n + nH_2O \xrightarrow{\text{enzyme}} nC_6H_{12}O_6$$

Starch,   Monosaccharides
a polysaccharide

Which of the following formulas represents a disaccharide?

— $C_5H_{10}O_5$ — $C_{12}H_{22}O_{11}$
— $(C_6H_{10}O_5)_n$ — $C_6H_{12}O_6$

$(C_6H_{10}O_5)_n; C_5H_{10}O_5$
and $C_6H_{12}O_6$

432

Which of the formulas listed in problem 431 represents a poly-saccharide? _____

Which are monosaccharides? _____

a. glucose, fructose, galactose
b. lactose, maltose, sucrose
c. starch, cellulose, glycogen

433

Other examples of polysaccharides are *cellulose*, found in plants, and *glycogen*, found in animals. Plants use the polysaccharide cellulose primarily for support. Animals store carbohydrate in the form of a polysaccharide, glycogen.

a. Name three monosaccharides. _____

b. Name three common disaccharides. _____

c. Name three polysaccharides. _____

434

Disaccharides are held together by a glycosidic bond. Polysaccharides are held together by _____ bonds.

435

a. Hydrolysis of a disaccharide yields _____.

b. Hydrolysis of a polysaccharide yields _____.

PROPERTIES
OF CARBOHYDRATES

Monosaccharides are crystalline compounds, soluble in water, and sweet to the taste, and need no digestion to be absorbed into the bloodstream.

Disaccharides are crystalline, water soluble, and sweet to the taste, and must be hydrolyzed (digested) to monosaccharides before they can be absorbed and used by the body for energy.

Polysaccharides are not water soluble and are not crystalline. They form colloidal dispersions instead of solutions. They are not sweet and must be hydrolyzed before being absorbed. (Cellulose is not hydrolyzed by enzymes in the human gastrointestinal tract.)

many glycosidic

a. **two monosaccharides**
b. **many mono-saccharides**

a. mono- and disaccharides
b. polysaccharides
c. mono- and disaccharides
d. polysaccharides
e. mono- and disaccharides
f. polysaccharides
g. monosaccharides

436

a. Which type(s) of carbohydrate is(are) soluble in water? _____

b. Which type(s) of carbohydrate form(s) a colloidal suspension in water? _____

c. Which type(s) of carbohydrate is(are) crystalline? _____

d. Which is(are) not crystalline? _____

e. Which is(are) sweet? _____

f. Which is(are) not sweet? _____

g. Which require(s) no digestion? _____

14

Lipids include such substances as fats, phospholipids, and steroids. Lipids are insoluble in water. Lipids contain the same elements as carbohydrates, but in lipids the proportion of oxygen is much lower than in carbohydrates.

__437__

Lipids contain the elements _____, _____, and _____.

carbon; hydrogen; oxygen

FATS

__438__

Fats are formed by the reaction of an organic acid with an alcohol. In fats, the organic acid is called a *fatty acid* and the alcohol is *glycerol* (sometimes called glycerin). The reaction is

$$\text{fatty acids} + \text{glycerol} \xrightarrow{\text{enzymes}} \text{fat} + \text{water}$$

The reverse reaction is the hydrolysis of a fat.

Thus, the hydrolysis of a fat yields _____ + _____.

fatty acids; glycerol

__439__

Fatty acids and glycerol are products of the hydrolysis of a _____.

fat

fatty acid; glycerol

440

The chemical equation for the hydrolysis of a fat may be written as

$$H-\underset{\underset{\displaystyle H}{\overset{\displaystyle H}{|}}}{\overset{\underset{\displaystyle |}{}}{C}}-O-\overset{\overset{\displaystyle O}{\|}}{C}-C_{17}H_{35}$$

A   +   3H_2O   → $3C_{17}H_{35}COOH$ +   B   C

Substance A is a fat.

Substance B is a _____.

Substance C is called _____.

a. contains carboxyl
    (—COOH) groups
b. contains hydroxyl
    (—OH) groups

441

a. How do you know that substance B is an acid? _____

b. How do you know that substance C is an alcohol? _____

a. many mono-
    saccharides
b. two monosaccharides
c. fatty acids and
    glycerol

442

a. The hydrolysis of a polysaccharide yields _____.

b. The hydrolysis of a disaccharide yields _____.

c. The hydrolysis of a fat yields _____.

443

Fatty acids that contain double bonds are called *unsaturated*. Unsaturated fatty acids tend to be more liquid than saturated fatty acids. A liquid fat is called an *oil*.

Increasing the number of double bonds between carbon atoms in fatty acids of a fat molecule should make the fat more _____ (solid/liquid). Decreasing the number of double bonds in a fat will make the fat more _____ (solid/liquid).

444

Which are unsaturated, fats or oils? _____

445

Unsaturated compounds contain double bonds. What kind of compounds would be called polyunsaturated? _____

446

Three fatty acids, butyric (C_3H_7COOH), palmitic ($C_{15}H_{31}COOH$), and stearic ($C_{17}H_{35}COOH$) acid, contain only single bonds. When they are present in a lipid, will that lipid tend to become solid or liquid? _____

447

Arachidonic acid is unsaturated, which means that it contains _____ (only single bonds/some double bonds) in its structure.

---

liquid; solid

oils

those containing many double bonds

solid

some double bonds

an oil; the fatty acid is unsaturated

448

A lipid that undergoes hydrolysis yields arachidonic acid and glycerol. Was the lipid a fat or an oil? _____

How do you know? _____

nonpolar

449

Lipids are insoluble in water. They are soluble in organic solvents such as gasoline, ether, carbon tetrachloride, and benzene. Organic solvents are nonpolar. Recall that, in general, polar compounds dissolve in polar liquids and nonpolar compounds dissolve in nonpolar liquids. Are lipids polar or nonpolar? _____

oleic acid is a *cis*-fatty acid; elaidic acid is a *trans*-fatty acid

450

Consider the following two fatty acids, oleic acid and elaidic acid.
Which is a *cis*-fatty acid? _____
Which is a *trans*-fatty acid? _____
(If you are not sure, see problem 392.)

Oleic acid

Elaidic acid

PHOSPHOLIPIDS

451

The hydrolysis of a phospholipid may be written as

phospholipid $\xrightarrow[\text{enzymes}]{\text{hydrolysis}}$ fatty acids + glycerol
+ phosphoric acid + an N-compound

The hydrolysis of a fat yields _____ + _____.

fatty acids; glycerol

452

The N-compounds in phospholipids are a certain group of nitrogen-containing compounds. Among them are serine, choline, inositol, and ethanolamine. Which is present determines the type of phospholipid, as will be discussed in the following problems.

The hydrolysis of a phospholipid yields _____

_____.

fatty acids, glycerol, phosphoric acid, and a nitrogen compound

453

a. Lipids contain which elements? _____

b. Phospholipids contain which elements? _____

a. C, H, O
b. C, H, O, P, N

454

Lecithin is a phospholipid in which the N-compound is *choline.*
Lecithin is a main component of cell membranes and is involved in the transportation of fats in the body. The hydrolysis of lecithin yields fatty acids, glycerol, phosphoric acid, and _____.

choline

fatty acids; glycerol;
phosphoric acid;
ethanolamine

455

Cephalin is a phospholipid involved in the blood-clotting process.
In cephalin, the N-compound is *ethanolamine.*

cephalin $\xrightarrow[\text{enzymes}]{\text{hydrolysis}}$ _____, _____,

_____, and _____.

STEROIDS

a. insoluble
b. insoluble
c. soluble

456

Another group, called steroids, is classified along with lipids because of
similar solubility characteristics.

a.   Lipids are _____ (soluble/insoluble) in water.

b.   Steroids are _____ (soluble/insoluble) in water.

c.   Steroids are _____ (soluble/insoluble) in organic liquids.

a. no
b. no

457

Steroids are high-molecular-weight tetracyclic (4-ring) compounds
whose structure is similar to

a.   Do steroids contain fatty acids? _____

b.   Do steroids contain glycerol? _____

458

Examples of steroids in the body are cholesterol, bile salts, sex hormones, and hormones of the adrenal cortex.

a.   Does cholesterol have a steroid structure? _____

b.   The ending *ol* in cholesterol indicates that it is what type of compound? _____

It contains which functional group? _____

a. yes
b. an alcohol; hydroxyl

Proteins 15

FORMATION AND HYDROLYSIS

459

Proteins contain the same elements as do carbohydrates and fats, except that proteins always contain one additional element: nitrogen.

All proteins contain the four elements: _____.

460

Some proteins also contain additional elements, such as sulfur, phosphorus, and iron. The hydrolysis of proteins yields amino acids. Likewise, the combination of amino acids yields proteins.

$$\text{protein} \; \underset{\text{combination}}{\overset{\text{hydrolysis}}{\rightleftharpoons}} \; \text{amino acids}$$

Amino acids contain what functional groups? _____

— COOH or carboxyl

461

The structure of an amino acid can be represented as

$$
\begin{array}{c}
H \quad\quad H \quad O \\
\backslash \quad\quad\; | \quad\quad \| \\
N - C - C - OH \quad \text{or} \quad H_2N - CH - COOH \\
/ \quad\quad | \quad\quad\quad\quad\quad\quad\quad\quad\quad | \\
H \quad\quad R \quad\quad\quad\quad\quad\quad\quad\quad\; R
\end{array}
$$

where "R" represents either a hydrogen or carbon-hydrogen groups.

Amino acids (and also proteins) can act as buffers because the —NH_2 group acts as a H^+ acceptor and the _____ group acts as a H^+ donor.

a. 3
b. many (or more than 3)

462

When two amino acids combine, the product is called a *dipeptide*.

a. A *tripeptide* would be formed when _____ amino acids combine.

b. A *polypeptide* would be formed when _____ amino acids combine.

amino acids

463

Note the formulas for alanine and glycine:

$$
\begin{array}{cc}
CH_3CHCOOH \quad\quad\quad\quad\quad CH_2COOH \\
| \quad\quad\quad\quad\quad\quad\quad\quad\quad\quad | \\
NH_2 \quad\quad\quad\quad\quad\quad\quad\quad NH_2 \\
\text{Alanine} \quad\quad\quad\quad\quad\quad \text{Glycine}
\end{array}
$$

Both would be classified as what type of compound? _____

464

When these two amino acids react, the following reactions are possible.

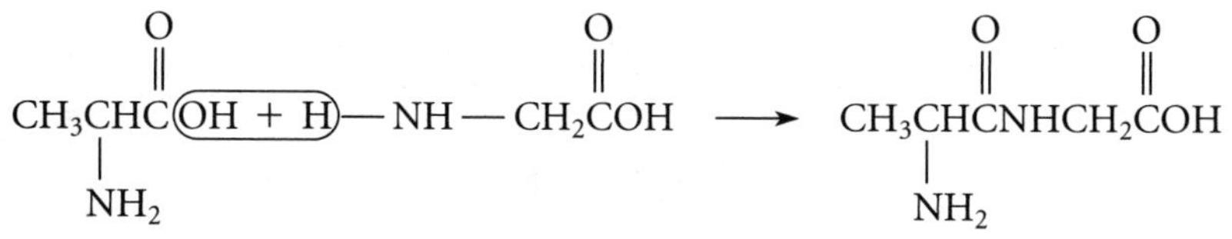

$$\underset{\text{Alanine}}{\underset{\underset{NH_2}{|}}{CH_3CHC}\underset{}{\overset{O}{\|}}(OH + H)-NH-\underset{\text{Glycine}}{CH_2C}\overset{O}{\|}OH} \longrightarrow \underset{\text{Alanyl-glycine (ala-gly)}}{\underset{\underset{NH_2}{|}}{CH_3CHC}\overset{O}{\|}NH\underset{}{CH_2C}\overset{O}{\|}OH}$$

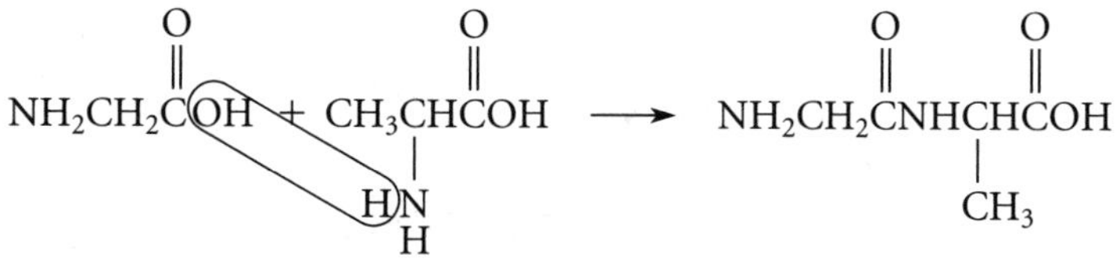

$$\underset{\text{Glycine}}{NH_2CH_2C\overset{O}{\|}(OH} + \underset{\text{Alanine}}{\underset{\underset{\underset{H}{|}}{HN}}{CH_3CHC}\overset{O}{\|}OH} \longrightarrow \underset{\text{Glycyl-alanine (gly-ala)}}{NH_2CH_2C\overset{O}{\|}NH\underset{\underset{CH_3}{|}}{CHC}\overset{O}{\|}OH}$$

Note that in both of these equations, the reaction is between the OH of a(n) _____ group and the H of a(n) _____ group.

carboxyl; amino

465

When glycine and alanine react, what type of compound is formed?

_____

a dipeptide

466

a. The abbreviation *ala* represents _____.

b. The abbreviation *ala-gly* represents _____.

a. alanine
b. alanyl-glycine

a tripeptide

467

If the amino acids glycine, alanine, and valine were combined in one compound, what type would it be? _____

ala-gly-val

468

What would be the abbreviation for a tripeptide containing alanine, glycine, and valine, in that order? _____

a tripeptide containing glycine, valine, and alanine, combined in that order

469

What would the abbreviation *gly-val-ala* represent? _____

_____

many; peptide

470

A dipeptide consists of two amino acids held together by a *peptide bond*. A tripeptide consists of three amino acids held together by peptide bonds. A polypeptide consists of _____ amino acids held together by _____ bonds.

carboxyl; amino

471

In a peptide bond (see problem 464), the bond occurs between a(n) _____ group of one amino acid and a(n) _____ group of another amino acid.

STRUCTURE

The *primary structure* of a protein refers to the linear sequence of amino acids in that protein. An example of a primary structure is

$$NH_2-CH_2-\overset{\overset{O}{\|}}{C}-NH-\underset{\underset{CH_3}{|}}{CH}-\overset{\overset{O}{\|}}{C}-NH-\underset{\underset{\underset{CH_3}{|}}{CH_2}}{CH}-\overset{\overset{O}{\|}}{C}-OH$$

Amino acid   Amino acid   Amino acid

$$\boxed{472}$$

What type of bond holds the amino acids together in the primary structure? _____

Note that all proteins have an amino ($-NH_2$) terminus and a carboxyl ($-COOH$) terminus.

peptide

The *secondary structure* of a protein refers to the regular recurring arrangement of the amino acid chain (its primary structure). One such arrangement, called the α-helix, occurs when the amino acid chain forms a spiral or coil, as shown here.

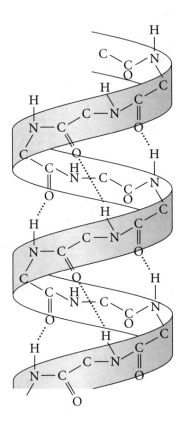

473

hydrogen

The coil consists of loops of the linear arrangement of the amino acids. These loops are held together by bonds between the O of the $C=O$ of one amino acid and the H of the NH of another amino acid.

Such a bond is called a _____ bond (refer to problems 379–383 if you don't know).

474

Another type of secondary structure, called the β-pleated sheet, consists of parallel strands of polypeptides, as shown in the following figure.

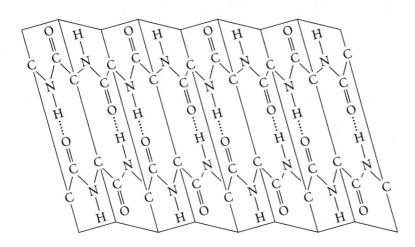

What type of bonds holds these sheets together? _____

475

a. The linear sequence of amino acids in a protein is called the _____ structure of that protein.

b. The regular recurring arrangement of the amino acid chains is called the _____ structure of a protein.

a. *α*-helix; *β*-pleated
   sheet
b. hydrogen

476

a. Two types of secondary structures are possible. They are the
   _____ and the _____.

b. What type of bond holds these secondary structures together?
   _____

The *tertiary structure* of a protein refers to the specific folding and bend-ing of the coils into layers or fibers, as shown here. It is this tertiary structure that gives proteins (and enzymes, which are also proteins) their biological activity.

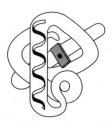

Some proteins have a *quaternary structure* that occurs when several pro-tein units, each with its own primary, secondary, and tertiary structure, combine to form a more complex unit, as shown in figure e. An exam-ple of a protein with a quaternary structure is hemoglobin.

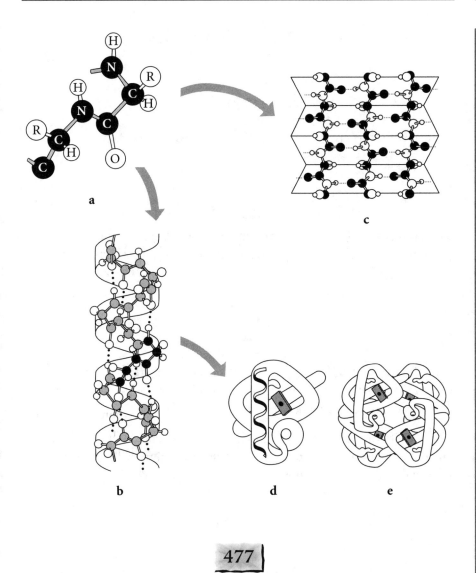

a

c

b   d   e

477

Look at the preceding five structures and label each as to primary, secondary, tertiary, or quaternary structure.

Which structure represents an α-helix? _____

Which structure represents a β-pleated sheet? _____

What type of bond holds the primary structures together? _____

What type of bond holds the secondary structure together? _____

a = primary;
b = secondary;
c = secondary;
d = tertiary;
e = quaternary

b; c; peptide; hydrogen

DENATURATION

weak

478

Hydrogen bonds are _____ (strong/weak) bonds.

no

479

Therefore, hydrogen bonds are easily broken. If the hydrogen bonds in a protein are broken, can the protein easily maintain its structural shape?

_____

cannot

480

If the hydrogen bonds in a protein are broken, the shape of the protein can change, making it incapable of performing its physiologic function. The protein is said to be *denatured*.

The activity of many proteins (and enzymes) depends on the presence of an active site that "fits" into a specific substrate. In figure **a**, the active site is shown in the center of the structure.

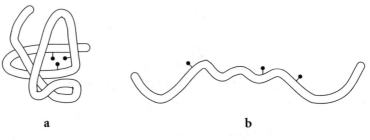

a   b

Enzymes

If the protein is denatured, the parts of the active site are no longer in close proximity (figure **b**), can no longer "fit" into the substrate, and therefore _____ (can/cannot) react.

481

a. hydrogen
b. reversible
c. irreversible denaturation

If the hydrogen bonds in a protein can easily re-form, then the denaturation is said to be *reversible*. If the hydrogen bonds cannot re-form, then the process is termed *irreversible denaturation*.

a.   In the denaturation of protein, _____ bonds are broken.

b.   If the broken hydrogen bonds can easily re-form, the process is called _____ denaturation.

c.   If the broken hydrogen bonds cannot re-form, the process is called _____.

Factors that can denature proteins include strong acids, heat, alcohol, and salts of heavy metals.

Nucleotides 16

ADENOSINE TRIPHOSPHATE

482

Nucleotides are formed by the reaction of a pentose (labeled Ⓢ for sugar), phosphoric acid (labeled Ⓟ), and a nitrogen-containing base (labeled Ⓝ), as shown:

$$Ⓝ—Ⓢ—Ⓟ$$

One example of a nucleotide is adenosine monophosphate, AMP, which is formed by the reaction of adenine (a nitrogen-containing base), ribose (a pentose), and phosphoric acid. From the name *adenosine monophosphate*, how many phosphoric acid molecules are present? _____

483

The structure of adenosine monophosphate, AMP, is shown in the following figure. It can be abbreviated as adenosine-P.

ribose; adenine

AMP

AMP contains which pentose? _____

Which nitrogen-containing base? _____

484

3

If a second molecule of phosphoric acid is bonded to the first phosphoric acid group, adenosine diphosphate, ADP, is formed. In adenosine triphosphate, ATP, how many phosphoric acid groups are present? _____

485

energy

ATP provides a form of chemical energy that is usable by all body cells. The structure of ATP may be abbreviated as

$$\text{adenosine} - \text{(P)} \sim \text{(P)} \sim \text{(P)}$$

Note that ATP contains two high-energy phosphate bonds indicated as $\sim$.

In the reaction

$$\text{ATP} \longrightarrow \text{ADP} + \text{(P)} + ? \qquad \text{or}$$

$$\text{adenosine} - \text{(P)} \sim \text{(P)} \sim \text{(P)} \longrightarrow \text{adenosine} - \text{(P)} \sim \text{(P)} + \text{(P)} + ?$$

what is produced other than ADP and P? _____

486

When ADP is converted into AMP, is energy released? _____

yes

487

To convert AMP back to ATP, what must be added? _____

energy and two phosphoric acid groups

488

Much of the energy required to form ATP comes from the metabolism of glucose.

a. AMP contains _____ phosphoric acid group(s) and _____ high-energy phosphate bond(s).

b. ADP contains _____ phosphoric acid group(s) and _____ high-energy phosphate bond(s).

c. ATP contains _____ phosphoric acid group(s) and _____ high-energy phosphate bond(s).

a. 1; 0
b. 2; 1
c. 3; 2

489

When two nucleotides are joined together, a dinucleotide is formed. One such dinucleotide is nicotinamide adenine dinucleotide, NAD.

a. Both nucleotides that make up NAD must contain which pentose?

_____

b. One nucleotide contains nicotinamide as the nitrogen-containing base (N-base); the other contains _____ as the nitrogen-containing base.

a. ribose
b. adenine

flavin, adenine

<u>490</u>

Which nitrogen-containing bases are present in flavin adenine dinucleotide, FAD? _____

Both dinucleotides, FAD and NAD, are involved in oxidation-reduction reactions in cells.

NUCLEIC ACIDS: DNA AND RNA

Nucleic acids are polymers of nucleotides. Consider the following structure of deoxyribonucleic acid, DNA, a nucleic acid. Note that each boxed segment is a nucleotide.

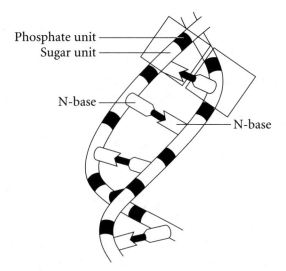

Phosphate unit
Sugar unit

N-base

N-base

DNA consists of a double coil with the opposite sides held together by hydrogen bonds. The backbones of the coils are alternating sugar and phosphate groups, with the hydrogen bonding between the nitrogen-containing bases holding the sides together.

491

In deoxyribonucleic acid, which pentose is a part of the coil backbone?

_____

deoxyribose

492

In DNA, there are four N-bases: adenine, guanine, thymine, and cytosine. They are abbreviated by the first letter (capitalized) of their name.

Adenine is abbreviated as _____.

Guanine is abbreviated as _____.

Thymine is abbreviated as _____.

Cytosine is abbreviated as _____.

A; G; T; C

493

Which N-base is represented by each of the following abbreviations?

T _____

A _____

C _____

G _____

thymine; adenine; cytosine; guanine

2; 3

494

In DNA, adenine is always hydrogen-bonded to thymine, and cytosine is always hydrogen-bonded to guanine. Thus, the structure of DNA may be represented as

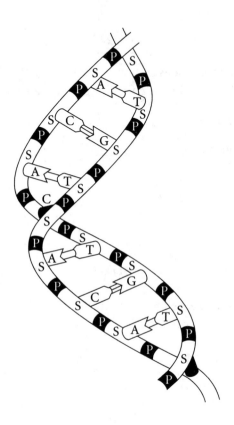

How many hydrogen bonds hold A and T together? _____

How many hydrogen bonds hold C and G together? _____

adenine, A; thymine, T; guanine, G; and cytosine, C

495

The four N-bases in DNA are _____

_____.

496

In DNA, adenine is always hydrogen-bonded to _____, and guanine is always bonded to _____.

thymine; cytosine

497

Complete the structure of the following figure, indicating missing N-bases and hydrogen bonds.

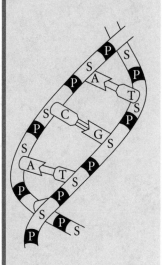

498

A second nucleic acid is ribonucleic acid, RNA. In RNA, the pentose is ribose; in DNA, the pentose is _____.

deoxyribose

499

In RNA, A is always bonded to uracil, U. In DNA, A is always bonded to _____.

T

a. G; 3
b. G; 3
c. T; 2
d. U; 2
e. **deoxyribose**
f. **ribose**

500

a. In DNA, C bonds to _____ with _____ (how many?) hydrogen bonds.

b. In RNA, C bonds to _____ with _____ (how many?) hydrogen bonds.

c. In DNA, A bonds to _____ with _____ (how many?) hydrogen bonds.

d. In RNA, A bonds to _____ with _____ (how many?) hydrogen bonds.

e. In DNA, the sugar is _____.

f. In RNA, the sugar is _____.

REPLICATION OF DNA

501

Consider the following illustration in which a DNA segment has been straightened out and the S's (sugars) and P's (phosphates) understood and not printed.

```
     T   C   A   A   G
     ‖   ‖‖‖ ‖   ‖   ‖‖‖
     A   G   T   T   C
```

If the DNA strands are separated, then each strand can act as a template to attract additional N-bases.

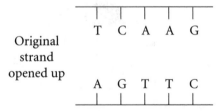

Original
strand
opened up

Let us consider the T and A at the far left of the open strand. The T will attract an A and the A will attract a T to begin the formation of a new backbone of a DNA molecule, as follows.

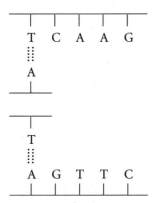

Now complete the balance of the strands for each DNA molecule, again showing N-bases and hydrogen bonding.

502

Are the two new DNA strands the same as the original one?

_____

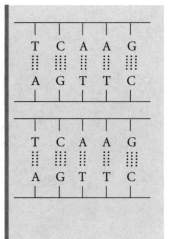

yes

TRANSCRIPTION OF INFORMATION (PROTEIN SYNTHESIS)

503

During transcription, DNA "unzips" and only part of one segment acts as a template for the formation of messenger RNA (mRNA).

In DNA, the partner of adenine (A) is _____.

In mRNA, the partner of adenine (A) is uracil, or _____.

T, or thymine; U

2; U; 2

In DNA, A is bonded to T with how many hydrogen bonds?

_____

Likewise in mRNA, A is bonded to _____ with how many hydrogen bonds? _____

deoxyribose; ribose

504

In DNA, the sugar is _____.

In mRNA, the sugar is _____.

505

Complete the following diagram indicating the partners and bonds needed for the formation of mRNA from the indicated DNA segment.

Segment of DNA | | | | | |
        C   A   G   T   A   C

mRNA   _____

(margin diagram:)
C   A   G   T   A   C
G   U   C   A   U   G

mRNA acts as a template for the transmission of genetic information from the cell's genes to the site of protein synthesis.

In the cytoplasm, another type of RNA is present. It is called transfer RNA or tRNA. The function of tRNA is to direct the production of the protein called for by mRNA.

tRNA has three N-bases at one end and an amino acid at the other end. It may be indicated by the following structure.

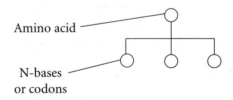

Amino acid

N-bases or codons

As mRNA travels across a ribosome, the first group of three N-containing bases picks up the corresponding tRNA. As the mRNA continues through the ribosome, the second group of three N-bases picks up its corresponding tRNA. The amino acid attached to the second tRNA bonds to the one attached to the first tRNA, and then the third one bonds to the second, and so on until a chain of amino acids (a protein) is formed, as in the following figure.

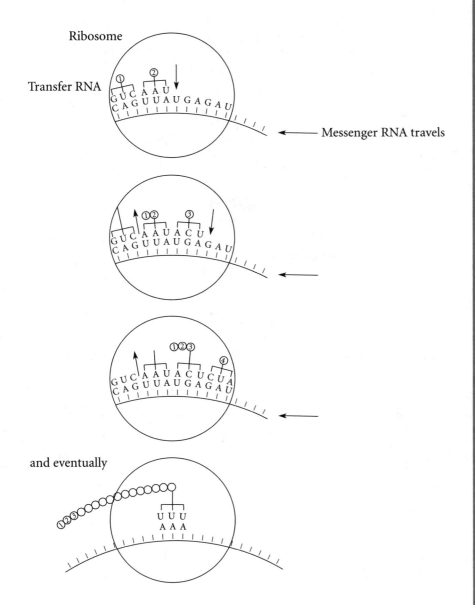

3

506

tRNA binds to mRNA at three consecutive sites, as in the following figure.

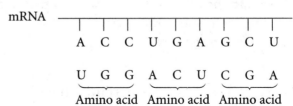

When tRNA binds to mRNA, how many bonding sites are there?

_____

These three binding sites are called the *triplet code* or *codon.*

507

In the diagram in problem 506, the beginning triplet code ACC corresponds to which code in tRNA? _____

UGG

508

Indicate the tRNAs and binding needed for the following segment of mRNA.

mRNA $\overline{\rule{3cm}{0pt}}$
       A C G C U U G C A U C C

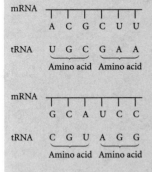

THE GENETIC CODE

Each tRNA is specific for a certain amino acid. The following chart indicates the amino acids corresponding to the triplet code in mRNA. Note that there is some redundancy; some amino acids are coded for more than one codon.

Messenger RNA

First Letter	Second Letter	Third Letter			
		A	C	G	U
A	A	lys	asn	lys	asn
	C	thr	thr	thr	thr
	G	arg	ser	arg	ser
	U	ile	ile	met*	ile
C	A	gln	his	gln	his
	C	pro	pro	pro	pro
	G	arg	arg	arg	arg
	U	leu	leu	leu	leu
G	A	glu	asp	glu	asp
	C	ala	ala	ala	ala
	G	gly	gly	gly	gly
	U	val	val	val	val
U	A	**	tyr	**	tyr
	C	ser	ser	ser	ser
	G	**	cys	trp	cys
	U	leu	phe	leu	phe

 *Chain initiator
**Chain terminator

509

Look at the answer in problem 508. The first tRNA bonds to the coded group ACG in mRNA. According to the preceding table, ACG in the mRNA corresponds to which amino acid? _____

thr

leu; ala; ser

510

The second tRNA in problem 508 corresponds to which amino acids?
_____ The third tRNA? _____ The fourth?
_____

The DNA segment above calls for four amino acids that are bonded in this order:

thr-leu-ala-ser

lys-arg-glu-ser

511

Indicate the amino acid chain designated by the following strand of mRNA.

| A | A | A | C | G | G | G | A | G | U | C | C |

_____

AUG; UAA, UAG, UGA

512

Which coded group(s) indicate the start of an amino acid chain?
_____ The termination of the chain? _____

ACC GCU UUU;
thr-ala-phe

513

Look at the following segment of DNA.

| T | G | G | C | G | A | A | A | A |

What is the sequence of mRNA for this segment: _____

What is the sequence of amino acids indicated? _____

MUTATIONS

514

Suppose that somehow one of the N-bases in the DNA in problem 513 was miscopied. Two such examples follow. In the first example, the sixth base is copied as a G instead of a U. In the second example, the last base is copied as an A instead of a U.

mRNA first miscopy: ACC GCG UUU

mRNA second miscopy: ACC GCU UUA

What sequence of amino acids would be indicated by the first miscopy? _____ By the second miscopy? _____

thr-ala-phe; thr-ala-leu

515

A change in copying DNA is called a mutation. Some mutations are not harmful; others might be. Would the change in the first miscopy cause a harmful mutation? _____ In the second miscopy? _____

no; maybe

GAA or GAG

516

The following diagram indicates the sequence of the 146 amino acids present in hemoglobin.

```
 1              5        10                  15
val - his - leu - thr - pro ┤ glu ├ glu - lys - ser - ala - val - thr - ala - leu - try - gly - lys ┐
                                                                                                    val
     30              25                  20
val - val - leu - leu - arg - gly - leu - ala - glu - gly - gly - val - glu - asp - val - asn ──────┘
35
tyr
     40              45                  50
pro - try - thr - gln - arg - phe - phe - glu - ser - phe - gly - asp - leu - ser - thr - pro ──────┐
                                                                                                    asp
         65              60                  55
     leu - val - lys - lys - gly - his - ala - lys - val - lys - pro - asn - gly - met - val - ala ─┘
gly
70              75                  80                  85
ala - phe - ser - asp - gly - leu - ala - his - leu - asp - asn - leu - lys - gly - thr - phe ──────┐
                                                                                                    ala
     100             95                  90
     asn - glu - pro - asp - val - his - leu - lys - asp - cys - his - leu - glu - ser - leu - thr ─┘
phe
     105             110                 115
arg - leu - leu - gly - asn - val - leu - val - cys - val - leu - ala - his - his - phe - gly ──────┐
                                                                                               120
                                                                                               lys
     135             130                 125
     val - gly - ala - val - val - lys - gln - tyr - ala - ala - gln - val - pro - pro - thr - phe - glu ─┘
ala
     140             146
     asn - ala - leu - ala - his - lys - tyr - his
```

Note that the sixth amino acid is glu. What is(are) the coded group(s) for glu? _____

GUA or GUG

517

Suppose that the middle section of either of these code groups is changed from A to U. Then the new coded groups would be either _____ or _____.

518

What amino acid(s) would this new coded group designate?

_____

val

519

Would this new protein of 146 amino acids still be hemoglobin?

_____

no

The new protein, with only one of 146 amino acids changed, is called hemoglobin-S and causes the genetic disease sickle-cell anemia.

Enzymes 17

Enzymes are proteins that act as biological catalysts. A catalyst increases the rate of a reaction but is not changed itself.

Pepsin is an enzyme found in gastric juice. It increases the rate of reaction for the digestion of protein in the stomach. Note that pepsin is an enzyme for a particular substance: protein. In general, enzymes are highly specific; catalysts are nonspecific.

520

Enzymes are _____ (specific/nonspecific) catalysts.

specific

521

a. What effect does an enzyme have on the rate of a reaction?
    _____

b. Is an enzyme changed by the reaction? _____

a. it increases it
b. no

NAMES OF ENZYMES

a. carbohydrates
b. lipids

522

Most enzymes have names ending in *ase*. Some enzymes are named for the type of reaction they catalyze. *Hydrolases* are enzymes that catalyze hydrolysis reactions. *Oxidases* are enzymes that catalyze oxidation reactions.

a.   Carbohydrases are enzymes that catalyze the hydrolysis of

_____.

b.   Lipases are enzymes that catalyze the hydrolysis of _____.

a. lactose
b. maltase

523

Other enzymes are named according to the substrate upon which they act. *Sucrase* is an enzyme that catalyzes the hydrolysis of sucrose.

a.   Lactase is an enzyme that catalyzes the hydrolysis of _____.

b.   Which enzyme catalyzes the hydrolysis of maltose? _____

COENZYMES

Some enzymes are proteins only. Other enzymes have a protein part and a nonprotein part, both of which must be present before the enzyme can function. The protein part of such an enzyme is called the *apoenzyme*; the nonprotein part is called the *coenzyme*. The reaction is

$$apoenzyme \ + \ coenzyme \ \longrightarrow \ enzyme$$

Some vitamins act as coenzymes. Nicotinamide adenine dinucleotide, NAD, acts as a coenzyme in oxidation-reduction reactions in the mitochondria. Coenzyme A (CoA) functions in the metabolism of carbohydrates, lipids, and proteins.

524

a. The nonprotein part of an enzyme is called the _____.

b. The protein part of an enzyme is called the _____.

a. coenzyme
b. apoenzyme

525

One example of a coenzyme is _____.

NAD or CoA or some vitamins

MODE OF ENZYME ACTIVITY

Enzymes are specific catalysts that contain an "active site," that section of the enzyme on which combination with the substrate takes place.

An older theory of enzyme activity, the *lock and key model*, assumes that the active site of an enzyme is rigid and that substrate molecules must fit into that site to react. Although it doesn't look like this illustration in reality, one small part of it might be represented like

Enzyme

If two compounds, A and B, approach the enzyme, they can attach themselves to it and then react together and go off as a compound, AB. The enzyme is left free to react again and is not used up.

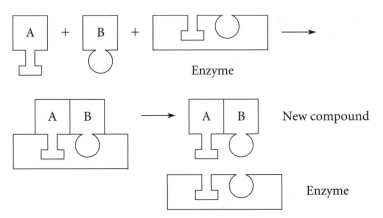

specific; increase

<u>526</u>

Enzymes are _____ (specific/nonspecific) catalysts that _____ the rate of chemical reaction in living systems.

Another version of the mode of enzyme activity is the *induced fit theory*. In this theory, the active site is flexible rather than rigid and it changes its conformation upon binding to the substrate, thus yielding an enzyme-substrate combination.

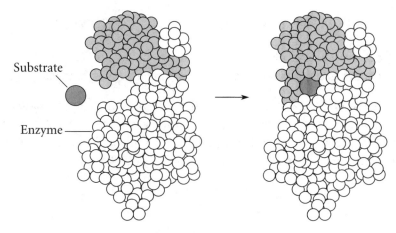

Induced fit model

a. rigid
b. flexible

<u>527</u>

a.   In the lock and key model, the active site is _____ (flexible/rigid).

b.   In the induced fit model, the active site is _____ (flexible/rigid).

INHIBITORS

528

If some substance other than the substrate fits into part of the active site, can the substrate then react with that enzyme? _____

no

529

A substance that competes for a position in the active site of an enzyme is called a *competitive inhibitor*. Which of the following substances could act as a competitive inhibitor for the active site shown in problem 530?

a

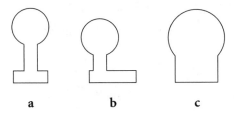

a   b   c

530

If substance **a** in problem 529 reacts with the active site of the enzyme shown here, could that enzyme then function normally? _____

no

Enzyme

it should inhibit
the growth of those
bacteria

<u>531</u>

An example of a competitive inhibitor is sulfanilamide, whose structure is similar to that of p-aminobenzoic acid, a compound that is essential for the growth of certain bacteria.

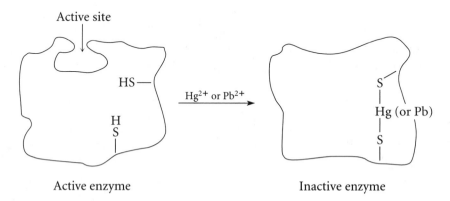

SO$_2$NH$_2$ COOH

NH$_2$ NH$_2$

Sulfanilamide   p-Aminobenzoic acid

What effect should sulfanilamide, a competitive inhibitor, have on the growth of the bacteria that need p-aminobenzoic acid? _____

no

<u>532</u>

Consider these diagrams:

Active site

HS —   S
                        |
Hg^{2+} or Pb^{2+} →   Hg (or Pb)
H |
S   S
| |

Active enzyme   Inactive enzyme

Note that the shape of the active site has been changed. Can the enzyme then function normally? _____

533

A substance that changes the shape of the active site by reacting with the enzyme at some point other than the active site is called a *noncompetitive inhibitor*. What substance acts as a noncompetitive inhibitor in the reaction in problem 532? _____

Hg^{2+} or Pb^{2+}

534

a.   A substance that fits into the active site of an enzyme and prevents that enzyme from functioning normally is called a(n) _____.

b.   A substance that changes the shape of the active site of an enzyme by reacting with that enzyme at a position other than the active site is called a(n) _____.

a. **competitive inhibitor**
b. **noncompetitive inhibitor**

Oxidation may be defined as

- a loss of an electron (or electrons)

- a gain of oxygen

- a loss of hydrogen

- an increase in oxidation number

The compound lactic acid may be oxidized to pyruvic acid according to the following equation.

$$H-\underset{\underset{H}{|}}{\overset{\overset{H}{|}}{C}}-\underset{\underset{(H)}{|}}{\overset{\overset{OH}{|}}{C}}\!\!-\!\!\overset{\overset{O}{\|}}{C}-OH \xrightarrow{\text{enzyme}} H-\underset{\underset{H}{|}}{\overset{\overset{H}{|}}{C}}-\overset{\overset{O}{\|}}{C}-\overset{\overset{O}{\|}}{C}-OH$$

Lactic acid   Pyruvic acid

535

This reaction is an oxidation reaction because it involves:

___ the gain of oxygen by the lactic acid molecule
___ the loss of hydrogen by the lactic acid molecule

the loss of hydrogen by the lactic acid molecule

a. hydroxyl and carboxyl
b. carbonyl and carboxyl

536

a. Lactic acid contains what functional groups? _____

b. Pyruvic acid contains what functional groups? _____

gain of oxygen

537

Another example of oxidation is the reaction of acetaldehyde to form acetic acid:

$$H-\underset{\underset{H}{|}}{\overset{\overset{H}{|}}{C}}-\overset{\overset{H}{|}}{C}=O \xrightarrow{[O]} H-\underset{\underset{H}{|}}{\overset{\overset{H}{|}}{C}}-\overset{\overset{OH}{|}}{C}=O$$

Acetaldehyde   Acetic acid

This reaction is oxidation because it involves _____.

there is a loss of an electron

538

The Fe^{2+} ion may be oxidized to the Fe^{3+} ion as follows:

$$Fe^{2+} - e^{-} \longrightarrow Fe^{3+}$$

Why is it oxidation? _____

Reduction is the opposite of oxidation. Reduction is

- ❧ a gain of an electron (or electrons)

- ❧ a gain of hydrogen

- ❧ a loss of oxygen

- ❧ a decrease in oxidation number

Oxidation and reduction always occur together. One cannot occur without the other.

539

Consider the following reaction:

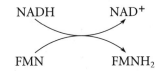

Succinic acid   Fumaric acid

a. Is succinic acid oxidized or reduced? _____
    Why? _____

b. Is FAD oxidized or reduced? _____ Why? _____

Note that both oxidation and reduction are involved in this reaction.

a. **oxidized; there is a**
    loss of hydrogen
b. **reduced; there is a**
    gain of hydrogen

540

Consider the following reaction:

$$NADH \qquad NAD^+$$

$$FMN \qquad FMNH_2$$

What was oxidized? _____ Why? _____

What was reduced? _____ Why? _____

NADH; it lost hydrogen;
FMN; it gained hydrogen

CELLULAR RESPIRATION

Cellular respiration enables cells to produce energy from food through oxidation-reduction reactions.

Look at Figure A in the Appendix and refer to it for problems 541–564. Note that cellular respiration involves three distinct stages: glycolysis, the Krebs cycle, and the electron transport chain.

cytoplasmic fluid; mitochondrion; mitochondrion

541

In which part of the cell does glycolysis take place? _____

In which part of the cell does the Krebs cycle occur?_____

In which part of the cell does the electron transport chain function? _____

pyruvic acid; cytoplasmic fluid

542

Glycolysis refers to the conversion of glucose to _____ and takes place in which part of the cell? _____

Krebs; mitochondrion

543

Pyruvic acid from the glycolysis sequence is changed to acetyl CoA and enters the _____ cycle, which takes place in the _____ of the cell.

NADH

544

What substance transfers electrons from the glycolysis sequence to the electron transport chain? _____

545

Where does the NADH from the glycolysis sequence go? _____

546

Where else is NADH produced? _____

547

During glycolysis, glucose is converted to _____.

548

Glycolysis takes place in the _____ of the cell.

549

In glycolysis, _____ is converted to pyruvic acid.

550

Electrons are transferred from the glycolysis sequence to the _____ via what substance? _____

551

The electron transport chain functions in which part of the cell?

_____

---

to the electron transport chain

in the Krebs cycle and in conversion of pyruvic acid to acetyl CoA

pyruvic acid

cytoplasmic fluid

glucose

electron transport chain; NADH

mitochondrion

mitochondrion

552

The Krebs cycle functions in which part of the cell? _____

3

553

Glycolysis produces pyruvic acid, NADH, and one other substance, ATP, the cell's main energy compound.

The abbreviation ATP refers to adenosine triphosphate. This name indicates that the compound adenosine is bonded to _____ (how many?) phosphate groups.

2

554

The abbreviation ADP refers to adenosine diphosphate, indicating that the adenosine molecule is bonded to _____ (how many?) phosphate groups.

adenosine; 1; phosphate

555

The abbreviation AMP refers to adenosine monophosphate and indicates that the _____ molecule is bonded to _____ (how many?) _____ group(s).

acetyl CoA

556

Pyruvic acid from the glycolysis sequence enters the Krebs cycle after being changed into _____.

557

Electrons are transported from the Krebs cycle via _____ and
_____.

NADH; FADH$_2$

558

Electrons are transported from the Krebs cycle to the _____.

electron transport chain

559

The electron transport chain functions in which part of the cell?

_____

mitochondrion

560

NADH is produced in three different sequences during cellular
respiration. They are _____, _____, and
_____.

glycolysis; conversion of
pyruvic acid to acetyl
CoA; Krebs cycle

561

FADH$_2$ is produced during which sequence? _____

Krebs cycle

562

The Krebs cycle produces the energy compound _____.

ATP

563

The electron transport chain produces the energy compound _____.

ATP

glycolysis; Krebs cycle;
electron transport chain

564

During cellular respiration, ATP is produced in the following sequence(s): _____, _____, and _____.

GLYCOLYSIS

Refer to Figure B in the Appendix for problems 565–584.

glucose; 6

565

Glycolysis begins with _____ (which compound?), which contains _____ (how many?) carbon atoms.

pyruvic acid; 3

566

The end product of glycolysis is _____, which contains _____ (how many?) carbon atoms.

2

567

For each molecule of glucose that goes through the glycolysis sequence, how many molecules of pyruvic acid are produced? _____

cytoplasmic fluid

568

In which part of the cell does glycolysis take place? _____

ADP

569

Note that in the first half of the glycolysis sequence ATP is used up and is converted to _____.

570

How many ATPs are required for the first three steps in the glycolysis sequence? _____

The ATP thus used is converted into _____.

2; ADP

571

In the second half of the glycolysis sequence, energy is produced and ADP is converted to _____.

ATP

572

In the second half of the glycolysis sequence, _____ (how many?) ADPs are converted to ATPs.

4

573

In the first half of the glycolysis sequence, _____ (how many?) ATPs are converted to ADPs.

2

574

In the entire glycolysis sequence, the net number of ATPs produced is _____.

2

2

575

In addition to ATP, another energy compound, NADH, is produced during glycolysis. How many NADHs are produced from one molecule of glucose during the glycolysis sequence? _____

NAD^+

576

NADH is produced from which compound? _____

pyruvic acid

577

The overall glycolysis sequence may be written as

glucose $\longrightarrow$ _____

2

578

The overall glycolysis sequence may be further written as

glucose $\longrightarrow$ _____ (how many?) pyruvic acid

ATP

579

A further addition to this equation is

$$\text{glucose} \xrightarrow[\hspace{2cm}]{\text{ADP} \qquad \underline{\hspace{3cm}}} \text{2 pyruvic acid}$$

580

This equation may then be modified to

$$\text{glucose} \xrightarrow[\text{2 ADP}]{\hspace{2em}\text{\_\_\_\_ (how many?) ATP}} \text{2 pyruvic acid}$$

581

This equation may be further modified to

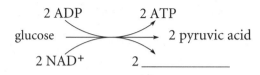

$$\text{glucose} \xrightarrow[\text{2 NAD}^+]{\text{2 ADP} \quad \text{2 ATP}} \text{2 pyruvic acid} \quad 2\underline{\hspace{4em}}$$

582

The complete glycolysis sequence may thus be written as

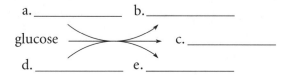

a._____ b._____

glucose

c._____

d._____ e._____

583

A biochemical reaction that requires oxygen is called *aerobic*; one that does not require oxygen is termed *anaerobic*.

Do any of the reactions in the glycolysis sequence require oxygen?

_____

anaerobic

584

Is glycolysis an aerobic or an anaerobic sequence? _____

CONVERSION OF PYRUVIC ACID TO ACETYL COENZYME A

Refer to Figure C in the Appendix for problems 585–598.

pyruvic acid

585

The end product of the glycolysis sequence is _____.

cytoplasmic fluid

586

Pyruvic acid is produced in which part of the cell? _____

mitochondrion

587

Pyruvic acid from the glycolysis sequence diffuses from the cytoplasmic fluid into which part of the cell? _____

no

588

Does pyruvic acid enter the Krebs cycle directly? _____

589

Pyruvic acid is converted into which compound in the mitochondrion?
_____

acetyl CoA

590

Does acetyl CoA (acetyl coenzyme A) enter the Krebs cycle? _____

yes

591

Each molecule of glucose that enters the glycolysis sequence produces
_____ (how many?) molecule(s) of pyruvic acid.

2

592

Each molecule of pyruvic acid that diffuses from the cytoplasmic
fluid into the mitochondrion produces _____ (how many?)
molecule(s) of acetyl CoA.

1

593

Therefore, during glycolysis, one molecule of glucose produces
_____ molecule(s) of pyruvic acid, which in turn are used
to produce _____ molecule(s) of acetyl CoA.

2; 2

6; 6; 2

594

One molecule of glucose contains _____ (how many?) carbon atoms.

Two molecules of pyruvic acid contain _____ (how many?) carbon atoms.

The acetyl groups of two molecules of acetyl CoA contain four carbon atoms.

How many carbon atoms are missing during this overall set of reactions? _____

2; pyruvic acid

595

As is shown in Figure C in the Appendix, in the cytoplasmic fluid, one molecule of carbon dioxide is produced for each molecule of pyruvic acid converted into acetyl CoA. Thus, one molecule of glucose enters the glycolysis sequence and is converted to _____ (how many?) molecule(s) of _____.

2; acetyl CoA

596

In the mitochondrion, two molecules of pyruvic acid are converted into _____ (how many?) molecule(s) of _____, which then enter the Krebs cycle.

carbon dioxide or CO_2

597

The missing carbon atoms show up as _____.

598

For each molecule of acetyl CoA produced from pyruvic acid, _____ (how many?) molecule(s) of NADH are produced. How many molecules of CO_2 are produced? _____

1; 1

THE KREBS CYCLE

Refer to Figure D in the Appendix for problems 599–609.

599

The two carbon atoms of the acetyl part of acetyl CoA participate in the Krebs cycle while the CoA molecule is recycled.

The two carbon atoms from the acetyl group are eliminated from the Krebs cycle as two molecules of what compound? _____

CO_2

600

As one molecule of acetyl CoA goes through the Krebs cycle:

How many molecules of ATP are produced? _____

How many molecules of NADH are produced? _____

How many molecules of $FADH_2$ are produced? _____

1; 3; 1

601

For every molecule of acetyl CoA that enters the Krebs cycle, _____ molecule(s) of ATP, _____ molecule(s) of NADH, and _____ molecule(s) of $FADH_2$ are produced.

1; 3; 1

ADP

NAD$^+$

FAD

mitochondrion

CO$_2$; ATP; NADH;
FADH$_2$

no

anaerobic

602

In the Krebs cycle, ATP is produced from what compound? _____

603

In the Krebs cycle, NADH is produced from what compound? _____

604

In the Krebs cycle, FADH$_2$ is produced from what compound? _____

605

In which part of the cell does the Krebs cycle take place? _____

606

The four products eliminated from the Krebs cycle are _____,
_____, _____, and _____.

607

Is oxygen required in any of the steps of the Krebs cycle? _____

608

Is the Krebs cycle an aerobic or anaerobic sequence? _____

609

Where do the NADH and the $FADH_2$ produced by the Krebs cycle go?

_____

to the electron
transport chain

ELECTRON TRANSPORT CHAIN

Refer to Figure E in the Appendix for problems 610–621.

610

The reactions in the electron transport chain take place in which part of
the cell? _____

mitochondrion

611

Is oxygen required in any of the steps of the electron transport chain?

_____

yes

612

Is the electron transport chain an aerobic or anerobic sequence?

_____

aerobic

613

Which compounds enter the electron transport chain? _____

NADH, $FADH_2$

614

The $FADH_2$ entering the electron transport chain comes from which
sequence? _____

Krebs cycle

glycolysis, conversion of pyruvic acid to acetyl CoA, and Krebs cycle

615

The NADH entering the electron transport chain comes from three different sequences. They are _____

_____.

ATP

616

What is the primary energy compound produced by the electron transport chain? _____

yes

617

Is ATP produced during glycolysis? _____

no

618

Is ATP produced during the conversion of pyruvic acid to acetyl CoA?

_____

yes

619

Is ATP produced in the Krebs cycle? _____

yes

620

Is ATP produced in the electron transport chain? _____

ADP

621

ATP is produced from which compound? _____

Photosynthesis 19

OVERALL REACTION

622

The overall reaction of photosynthesis is

$$6CO_2 + 6H_2O \xrightarrow[\text{chlorophyll}]{\text{light}} C_6H_{12}O_6 + 6O_2$$

During photosynthesis,

a. the raw materials are _____

b. the end products are _____

c. the source of energy is _____

d. the source of CO_2 is _____

e. the source of H_2O is _____

f. what happens to the O_2? _____

a. CO_2 and H_2O
b. $C_6H_{12}O_6$ and O_2
c. light
d. air
e. the ground
f. released into air

animal; plant

623

Look at the following structures for heme (which is a part of the hemoglobin molecule) and chlorophyll *a*.

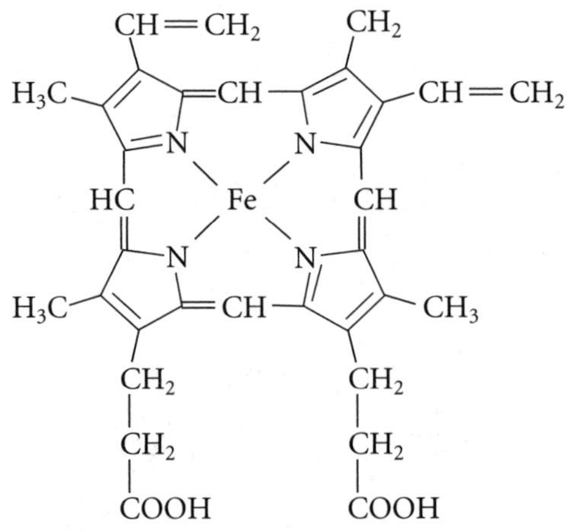

Heme

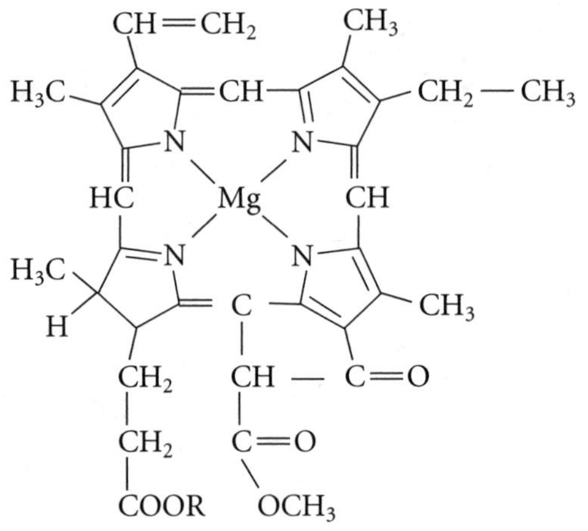

Chlorophyll *a*

Note that these two substances have similar structures. Heme (hemoglobin), however, is necessary for most _____ (animal/plant) life, whereas chlorophyll *a* is necessary for most _____ (animal/plant) life.

624

What metal is at the center of the heme molecule? _____ The chlorophyll *a* molecule? _____

625

Consider the following two examples of the overall photosynthesis reaction. Radioactive oxygen, ^{18}O, is introduced in the form of radioactive $C^{18}O_2$ in the first reaction and as radioactive water, $H_2^{18}O$, in the second reaction.

$$6C^{18}O_2 + 6H_2O \xrightarrow[\text{chlorophyll}]{\text{light}} C_6H_{12}{}^{18}O_6 + 6O_2$$

$$6CO_2 + 6H_2^{18}O \xrightarrow[\text{chlorophyll}]{\text{light}} C_6H_{12}O_6 + 6^{18}O_2$$

What is the source of the O_2 produced during photosynthesis, CO_2 or H_2O?

626

Photosynthesis may be subdivided into two main steps as indicated in the following simplified chart.

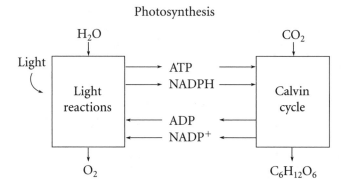

Photosynthesis

a. light
b. H_2O
c. ATP, NADPH; O_2
d. H_2O

For the first step, *light reactions*,

a.   what is the energy source? _____

b.   what reactant is necessary? _____

c.   what compounds are produced? _____

d.   where does the O_2 come from? _____

a. CO_2
b. carbohydrate or
   $C_6H_{12}O_6$
c. ADP; $NADP^+$
d. no

627

For the second step, the Calvin cycle,

a.   what is the principal reactant? _____

b.   what is the principal product? _____

c.   what energy compounds are returned to the light reactions step?

   _____

d.   Is light required for the Calvin cycle? _____

LIGHT REACTIONS

The following is a simplifed diagram of light reactions.

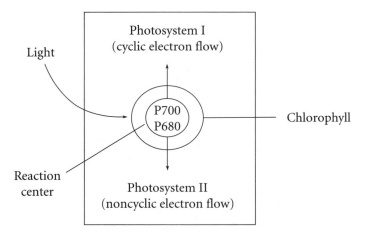

628

When light strikes the chlorophyll molecules, it excites an electron, which is transferred from chlorophyll molecule to chlorophyll molecule until it reaches the reaction center. In photosystem I, the reaction center is associated with a specific protein (pigment) known as P700 because it is most effective at absorbing light with a wavelength of 700 nm. Photosystem II is associated with a similar protein P680 that is sensitive to absorbing light with a _____.

wavelength of 680 nm

629

Which photosystem has a noncyclic electron flow? _____

II

P700; P680

630

What numbered form of chlorophyll *a* is associated with photosystem I? _____ With photosystem II? _____

I; II

631

Which photosystem is cyclic? _____ Noncyclic? _____

The following is an abbreviated diagram of photosystem I.

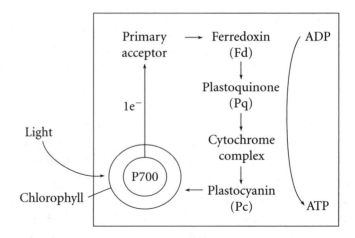

1

632

How many electrons are involved in this system? _____

yes

633

In photosystem I, an electron is emitted by P700 and travels to a primary acceptor. After a series of reactions, does the electron return to P700?

_____

That is why this system is called cyclic.

634

Is H_2O split to provide electrons in photosystem I? _____

no

635

Is O_2 produced during the reactions of photosystem I? _____

no

636

What product is the result of the reactions in photosystem I?

_____

ATP

637

In photosystem I, ATP is produced from _____.

ADP

638

When ADP is converted to ATP, what group must be added?

_____

P (or phosphate)

639

The production of ATP during the cyclic electron flow is called *cyclic photophosphorylation*.

During cyclic photophosphorylation, _____ is converted to _____.

ADP; ATP

no

640

Is NADPH produced in this sequence? _____

no

641

Is O_2 produced in this sequence? _____

P700; light

642

The energy for the electron involved in photosynthesis I is produced from the chlorophyll *a* that is labeled _____ and in turn is produced by the action of _____ on chlorophyll.

The following is an abbreviated diagram of photosystem II.

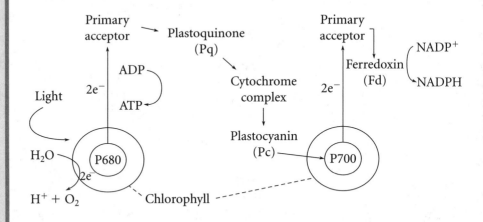

yes; no;
yes; yes

643

In photosystem I, is ATP produced? _____ NADPH?

_____

In photosystem II, is ATP produced? _____ NADPH?

_____

644

Where do the electrons used by photosystem II come from?

_____

645

When the H_2O gives off electrons, what substances are produced?

_____

646

Is the same type of chlorophyll _a_ needed for both photosystems?
(yes/no) _____ Why?

647

Can ATP be produced in both photosystems? _____

648

Is photosystem II cyclic? _____

649

Is water splitting required for photosystem I? _____ For photosystem II? _____

650

Is O_2 produced in photosystem I? _____ In photosystem II?

_____

water

H^+, O_2

no, system I needs P700
and system II needs P680

yes

no

no; yes

no; yes

no; yes

651

Is NADPH produced in photosystem I? _____ In photosystem II? _____

$NADP^+$

652

From what substance is NADPH produced? _____

H_2O

653

From what substance is O_2 produced? _____

The light reactions system may be abbreviated as shown in the following diagram.

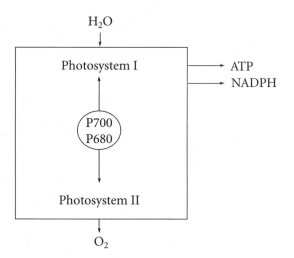

light

654

What is the original energy source for this series of reactions?

_____

655

When light strikes photosystem I and photosystem II, it causes electrons to flow.

Which of these systems is cyclic? ____________

Which is noncyclic? ____________

I; II

656

In which photosystem is water splitting involved? ____________

II

657

What is the source of oxygen during the light reactions? ____________

water

658

By which photosystem is oxygen produced? ____________

II

659

What high-energy compound(s) is(are) produced in photosystem I?

____

ATP

660

What high-energy compound(s) is(are) produced in photosystem II?

____

ATP, NADPH

ADP

NADP⁺

3

661

From what substances is ATP produced? _____

662

From what substance(s) is NADPH produced? _____

CALVIN CYCLE

The following is an abbreviated diagram of the Calvin cycle.

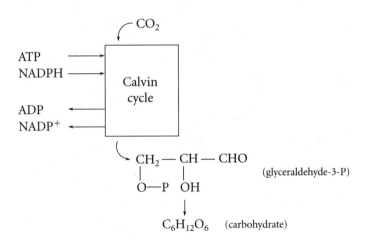

663

The end product of the Calvin cycle is glyceraldehyde-3-P. It contains how many carbon atoms? _____

664

CO_2 contains how many carbon atoms? _____

1

665

How many CO_2 molecules must go through the Calvin cycle to produce one molecule of glyceraldehyde-3-P? _____

3

666

Is CO_2 a part of the light reactions sequence or the Calvin cycle?

_____

Calvin cycle

667

What high-energy compounds enter the Calvin cycle? _____

ATP, NADPH

668

Where do these energy compounds come from? _____

light reactions

The following is a simplified illustration of the Calvin cycle. Numbers in brackets [   ] indicate total number of carbon atoms. Enzymes for each step are omitted for simplicity.

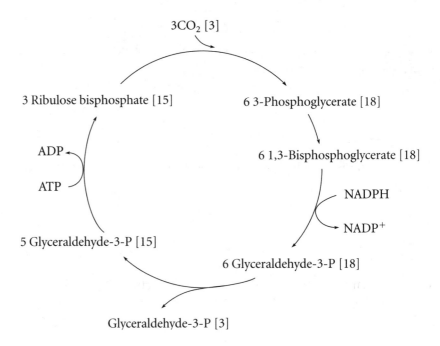

Simplified Calvin cycle

How many carbon atoms enter the cycle? _____

669

In what compound do they enter? _____

670

In what compound do the carbons exit the Calvin cycle? _____

671

3

CO_2

glyceraldehyde-3-P

672

Glyceraldehyde-3-P contains how many carbons? _____

3

673

Does all the glyceraldehyde-3-P produced in the Calvin cycle leave the cycle? _____

no

674

What fraction of the glyceraldehyde-3-P leaves that cycle?

_____

⅙

675

Does the number of carbon atoms entering, reacting in, and leaving the cycle balance? _____

yes

676

What energy compounds enter the Calvin cycle? _____

ATP, NADPH

677

What energy compounds are produced in the Calvin cycle?

_____

ADP, NADP⁺

ATP; NADPH

678

These energy compounds return to the light reactions system. There, ADP is regenerated to _____ and $NADP^+$ is regenerated to _____.

3

679

Glyceraldehyde-3-P contains how many carbon atoms? _____

6

680

Glucose, $C_6H_{12}O_6$, contains how many carbon atoms? _____

2

681

How many glyceraldehyde molecules must be combined to produce one glucose (carbohydrate) molecule? _____

The following figure gives an overall summary of photosynthesis.

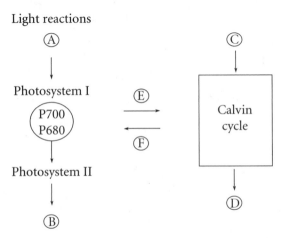

Summary of photosynthesis

682

What is the substance A that enters the light reactions system?

_____

683

What is substance B? _____

684

What compounds are represented by E? _____

685

What compounds are represented by F? _____

686

What is compound C? _____

687

What is compound D? _____

H_2O

O_2

ATP, NADPH

ADP, NADP$^+$

CO_2

glyceraldehyde-3-P

C₄ PLANTS

C_4 plants are those that have an alternate method of fixing CO_2 for the Calvin cycle as indicated in the following abbreviated diagram. (Numbers in brackets [   ] refer to the number of carbon atoms in each molecule.)

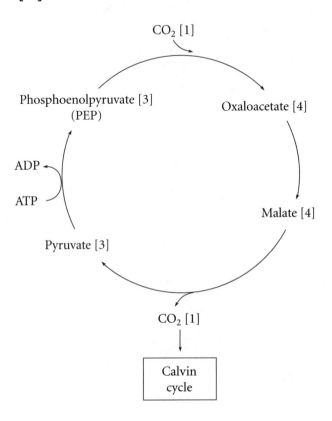

air

688

What is the source of CO_2? _____

malate, oxaloacetate

689

What four-carbon compound(s) (is/are) involved in the C_4 cycle?

_____

690

What high-energy compound is needed for the C$_4$ cycle? _____

691

When CO$_2$ enters the Calvin cycle, what high-energy compound(s) (is/are) required? _____

692

Where are these high-energy compounds produced? _____

ATP

ATP, NADPH

in light reactions
sequence

Oxygen–Carbon Dioxide Transport in Blood 20

FLOW OF GASES

Gases diffuse from an area of higher pressure to one of lower pressure.

In the following diagram, P_{O_2} indicates the partial pressure of oxygen (that part of the total pressure exerted by the oxygen) in millimeters of mercury.

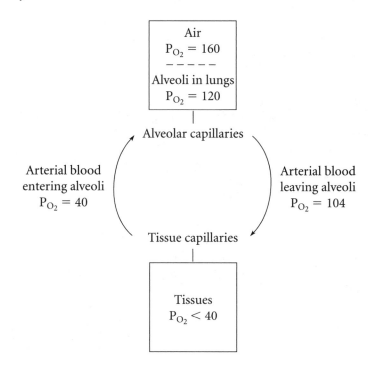

Air
$P_{O_2} = 160$
– – – – –
Alveoli in lungs
$P_{O_2} = 120$

Alveolar capillaries

Arterial blood
entering alveoli
$P_{O_2} = 40$

Arterial blood
leaving alveoli
$P_{O_2} = 104$

Tissue capillaries

Tissues
$P_{O_2} < 40$

into the capillaries in the alveoli

gases diffuse from higher to lower partial pressure

picks up

693

Which way does oxygen diffuse in the lungs? _____

694

From the alveoli, oxygen diffuses into the blood and then to the tissues. Why does it move in that direction? _____

695

When the arterial blood flows back to the alveoli, does it pick up or give off oxygen? _____

Tissues require oxygen for metabolic reactions such as

$$C_6H_{12}O_6 + 6O_2 \longrightarrow 6CO_2 + 6H_2O$$

Carbon dioxide is a gaseous waste product of oxidative metabolism. Its partial pressures in various parts of the blood system are shown in the following figure.

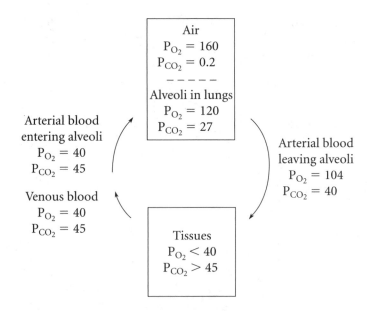

696

From the tissues, which way does carbon dioxide flow? _____

_____

697

Why does carbon dioxide flow in that direction? _____

_____

TRANSPORT OF OXYGEN

698

When oxygen enters the red blood cells, it binds with hemoglobin, here represented by the formula HHb, to form a compound known as oxy-hemoglobin.

$$HHb + O_2 \longrightarrow HHbO_2$$

Hemoglobin   Oxyhemoglobin

Oxyhemoglobin in turn is ionized into the oxyhemoglobin ion and a hydrogen ion.

$$HHbO_2 \longrightarrow H^+ + HbO_2^-$$

Oxygen is carried to the tissues in the form of the oxyhemoglobin ion, or _____.

699

In the tissues, the oxyhemoglobin ion binds with a hydrogen ion to form oxyhemoglobin; the formula is

$$HbO_2^- + H^+ \longrightarrow \text{_____}$$

---

by diffusion to venous blood to arterial blood to alveoli to lungs to air

gases flow from an area of higher partial pressure to one of lower partial pressure

HbO_2^-

$HHbO_2$

$$HHbO_2 \longrightarrow HHb + O_2$$

700

The oxyhemoglobin releases its oxygen to form hemoglobin. This reaction may be written as _____

_____

The oxygen thus liberated is used by the tissues for various metabolic reactions. The hemoglobin once again is available to pick up oxygen in the alveoli.

hemoglobin

701

What substance in the blood carries oxygen to the tissues? _____

HHb

702

What is the symbol for hemoglobin? _____

oxyhemoglobin

703

When hemoglobin picks up oxygen, _____ is formed.

$$HHbO_2$$

704

What is the formula of oxyhemoglobin? _____

$$HHb + O_2 \longrightarrow HHbO_2$$

705

Write the reaction for hemoglobin and oxygen. _____

_____

706

Write the formula for the ionization of oxyhemoglobin. _____

_____

$$HHbO_2 \rightarrow H^+ + HbO_2^-$$

707

In what form is oxygen carried to the tissues? _____

HbO_2^- or oxyhemoglobin ion

708

Write the reaction of the oxyhemoglobin ion with a hydrogen ion in the tissues. _____

_____

$$HbO_2^- + H^+ \rightarrow HHbO_2$$

709

Write the reaction indicating how oxyhemoglobin gives off its oxygen in the tissues. _____

_____

$$HHbO_2 \rightarrow HHb + O_2$$

710

What happens to the hemoglobin after it gives off its oxygen?

_____

returns to alveoli to pick up more O_2

red blood cells

What part of the blood contains hemoglobin? _____

The overall steps in this process may be diagrammed as follows.

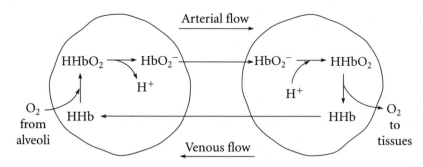

Red blood cells
in lung capillaries

Red blood cells
in tissues

TRANSPORT OF CARBON DIOXIDE

**gases diffuse from an
area of higher to lower
partial pressure**

Carbon dioxide is produced in the tissues and diffuses into the red blood cells.

Why does it diffuse in that direction? _____
_____

713

In the red blood cells, carbon dioxide reacts with water to produce carbonic acid:

$$CO_2 + H_2O \longrightarrow H_2CO_3$$
Carbonic acid

Carbonic acid in turn ionizes, producing a hydrogen ion and a bicarbonate ion.

$$H_2CO_3 \longrightarrow H^+ + HCO_3^-$$
Bicarbonate
ion

Red blood cells are able to hold only a small amount of the bicarbonate ions produced from carbon dioxide, so the excess diffuses into the blood plasma.

The bicarbonate ion has what charge? _____

714

To replace the bicarbonate ions, chloride ions (Cl^-) flow from the plasma into the red blood cells.

Chloride ions have what charge? _____

715

The process of shifting bicarbonate and chloride ions in opposite directions across a red blood cell membrane is called the *chloride shift*. It takes place in the red blood cells in the lung and tissue capillaries.

In the chloride shift, _____ ions flow from the red blood cells to the _____.

−1

−1

HCO_3^-; plasma

Cl⁻; plasma; red blood cells

$$H_2CO_3 \longrightarrow H_2O + CO_2$$

most of it is exhaled

chloride; bicarbonate

716

As bicarbonate ions flow out of the red blood cells, _____ ions flow from the _____ into the _____.

717

In the lungs, the bicarbonate ion combines with a hydrogen ion to re-form carbonic acid.

$$HCO_3^- + H^+ \longrightarrow H_2CO_3$$

Carbonic acid in turn decomposes into water and carbon dioxide, which is then exhaled. The equation for this reaction is _____.

718

What happens to the carbon dioxide in the lungs? _____

719

As the bicarbonate ion is used up in the preceding reactions, it is replaced by bicarbonate ions from the plasma. At the same time, chloride ions from the red blood cells flow back into the plasma. This process is termed a *reverse chloride shift*.

In a reverse chloride shift, _____ ions flow from the red blood cells into the plasma and at the same time _____ ions flow from the plasma into the red blood cells.

<u>**720**</u>

These reactions may be diagrammed as follows.

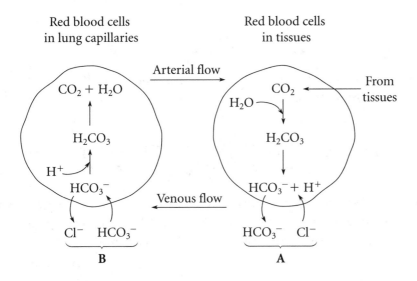

Red blood cells in lung capillaries

Red blood cells in tissues

What does part **A** represent? _____

What does part **B** represent? _____

<u>**721**</u>

In addition to the red blood cells transporting carbon dioxide as bicarbonate ions, some of the carbon dioxide reacts with hemoglobin to form carbaminohemoglobin:

$$CO_2 + HHb \longrightarrow HHbCO_2$$

In the lungs, carbaminohemoglobin decomposes into hemoglobin and carbon dioxide.

What happens to the carbon dioxide? _____

<u>**722**</u>

What happens to the hemoglobin? _____

A: chloride shift;
B: reverse chloride shift

it is exhaled

it is recycled

hemoglobin

bicarbonate; carbamino-hemoglobin

723

Oxygen is carried to the tissues by what compound? _____

724

Carbon dioxide is carried from the tissues in the form of _____ ions and also in the compound _____.

TRANSPORT OF BOTH OXYGEN AND CARBON DIOXIDE

The interrelationship of the oxygen and carbon dioxide transport systems is shown in the following diagram. Note how the hydrogen ions in the red blood cells in both the lungs and the tissues are recycled.

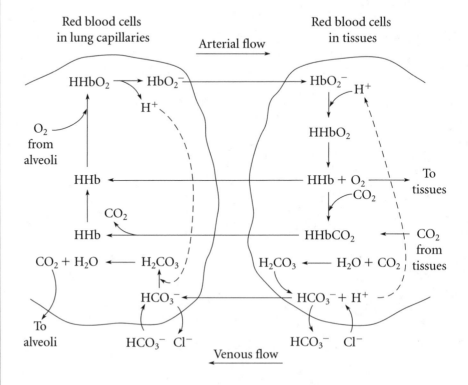

Review

725

What elements do the following symbols represent?

a. H _____   b. O _____   c. S _____

d. Na _____   e. K _____   f. Fe _____

726

Write the symbol for each of the following elements:

a. chlorine _____   b. zinc _____   c. calcium _____

d. carbon _____   e. bromine _____   f. magnesium _____

727

When an atom gains an electron, it forms an ion with what charge?

_____

728

The chlorine atom gains 1 electron to form a chloride ion. The symbol for the chloride ion is _____.

ionic

> **729**
>
> When two ions are held together by the attraction of their opposite charges, what type of bond is between them? _____

hydrogen (or H$^+$)

> **730**
>
> An acid is a substance that yields _____ ions in solution.

hydrogen (or H$^+$)

> **731**
>
> A base is a substance that accepts _____ ions in solution.

any ions except H$^+$ and OH$^-$

> **732**
>
> Salts yield what types of ions? _____

an electrolyte

> **733**
>
> A solution that conducts electricity is called _____.

a nonelectrolyte

> **734**
>
> A solution that does not conduct electricity is called _____.

735

Which of the following are anions? _____

Cations? _____

$$Na^+ \qquad Cl^- \qquad Mg^{2+}$$
$$SO_4^{2-} \qquad K^+ \qquad NO_3^-$$

anions: Cl^-, SO_4^{2-}, NO_3^-; cations: Na^+, Mg^{2+}, K^+

736

Saliva has a pH between 5.5 and 6.9. What type of liquid is saliva?

_____

a weak acid

737

Blood has a pH range of 7.35–7.45. What type of liquid is blood?

_____

a weak base

738

An acid of pH 2.75 is how many times as strong as one of pH 3.75?

_____

10

739

A solution that maintains a constant pH upon the addition of either acid or base is called a(n) _____.

buffer

sugar; water

740

In a solution of sugar in water, the solute is _____ and the solvent is _____.

yes; no

741

Do solutions pass through membranes? _____ Do colloids? _____

100

742

A 10% glucose solution will contain 10 g of glucose in _____ mL of solution.

1 liter

743

A 1M KCl solution will contain 1 mole of KCl in how much solution? _____

0.5M NaCl

744

Which solution will have a higher osmolarity, 0.5M NaCl or 0.5M glucose? _____

covalent

745

When two atoms share electrons, they are held together by a _____ bond.

solute

746

Diffusion is the flow of _____ (solute/solvent).

747

Osmosis is the flow of _____ .

water

748

Is osmosis active or passive transport? _____

passive

749

Is diffusion a form of active transport? _____

no

750

A hypertonic solution has a concentration _____ (higher than/ lower than/the same as) that of another solution.

higher than

751

An isotonic solution has a concentration _____ (higher than/ lower than/the same as) that of another solution.

the same as

752

A hypotonic solution has a concentration _____ (higher than/ lower than/the same as) that of another solution.

lower than

753

A selective membrane is _____ (permeable/semipermeable/ nonpermeable).

semipermeable

opposite directions

754

Do osmosis and diffusion normally take place in the same direction or in opposite directions? _____

K^+; Na^+

755

Which ion, Na^+ or K^+, has a higher intracellular concentration in animal cells? _____ Extracellular concentration? _____

ATP

756

What high-energy compound frequently supplies the energy for active transport? _____

Na^+out; K^+ in

757

During active transport by the Na/K pump, which ions are transported out of the cell? _____ Into the cell? _____

$3 Na^+$out; $2 K^+$ in

758

What are the relative numbers of Na^+ and K^+ ions transported out of and into a cell? _____

–

759

Because of active transport, the interior of the cell has a relative _____ $(+/-)$ charge.

760

When a nerve cell is resting, are its sodium channels normally open or closed? _____ Its potassium channels? _____

761

When a stimulus affects a resting nerve cell, which channels open?

_____

762

When a stimulus affects a resting nerve cell, which ions move, and do they enter or leave the cell? _____

763

When a typical resting nerve cell has a stimulus applied, it becomes _____ (polarized/depolarized).

764

After a nerve cell has been depolarized, the _____ channels close and the _____ channels open.

765

This latter change is called _____.

both closed

sodium

Na$^+$ enter

depolarized

Na$^+$; K$^+$

repolarization

K+

active transport

Ca²⁺

causes vesicles to fuse to presynaptic membrane

neurotransmitters; postsynaptic

they cause ion channels to open, which in turn brings about the depolarization of the postsynaptic membrane and thus the transmission of the impulse along the next axon

766

During the undershoot, which channels are slow to close?

_____

767

After the undershoot, which process returns the cell to its resting state?

_____

768

When a new impulse reaches the presynaptic membrane of an axon, it depolarizes that membrane, causing an influx of _____ ions.

769

The calcium ions have what effect on the vesicles near the presynaptic membrane? _____

770

When the vesicles fuse with the presynaptic membrane, they release _____, which travel across the synaptic cleft to the _____ membrane.

771

What effect do the neurotransmitters have on the postsynaptic membrane? _____

_____

_____

772

Diagram the structure of the hydrocarbon compound containing the following arrangement of carbon atoms, and indicate all the hydrogen atoms.

```
        C
        |
   C — C — C
        |
        C
```

```
          H
          |
      H — C — H
   H              H
   |              |
H — C —— C —— C — H
   |              |
   H              H
      H — C — H
          |
          H
```

773

Indicate the hydrogen atoms attached to the following arrangement of carbon atoms.

```
   C ≡ C — C
```

```
               H
               |
H — C ≡ C — C — H
               |
               H
```

774

What is the simplified structure of this compound?

```
    H   H   H   H   H
    |   |   |   |   |
H — C — C — C — C — C — H
    |   |   |   |   |
    H   |   H   H   H
        |
    H — C — H
        |
        H
```

$CH_3CH(CH_3)CH_2CH_2CH_3$

polar

hydrogen

hydrogen

775

What is the simplified structure of this compound?

776

Is the water molecule polar or nonpolar? _____

777

What types of bonds are present *between* water molecules? _____

778

What type of bond is involved in maintaining the coil shape of a protein? _____

779

When DNA replicates, it "unzips," breaking what type of bonds?
_____

hydrogen

780

Are hydrogen bonds strong or weak? _____

weak

781

In a protein, hydrogen bonds occur between an H attached to a(n)
_____ atom and an O that is part of a(n) _____
group.

N; C=O

782

What type of functional group is represented by the following?

a. OH _____   b. COOH _____

c. C—C—C _____   d. OPO$_3$H$_2$ _____
     ||
     O

e. NH$_2$ _____   f. CHO _____

a. hydroxyl
b. carboxyl
c. carbonyl
d. phosphate
e. amino
f. carbonyl

Identify the types of functional groups present in each of the compounds in problems 783–785.

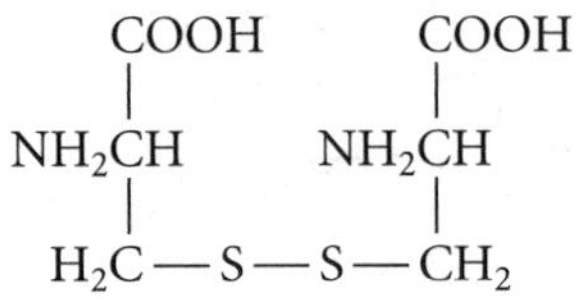

carboxyl, amino, disulfide

783

COOH   COOH
| |
NH₂CH   NH₂CH
| |
H₂C — S — S — CH₂

sulfhydryl

784

CH₃CH₂CH₂SH

a. amino and carbonyl
b. carbonyl and hydroxyl
c. hydroxl and carboxyl

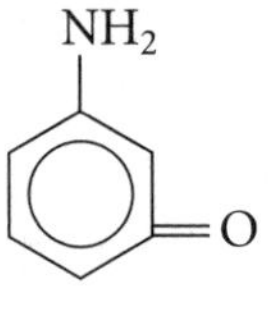

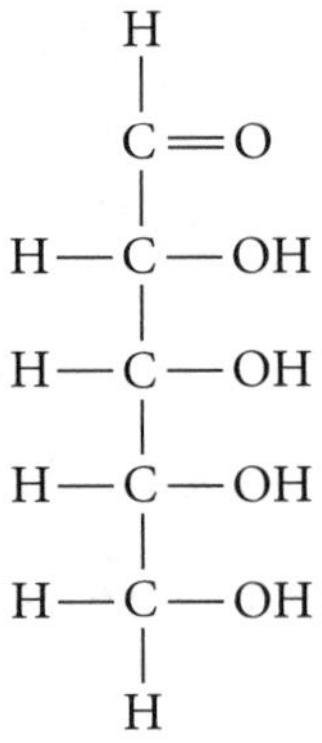

785

a.

NH₂

O

b.

H
|
C=O
|
H — C — OH
|
H — C — OH
|
H — C — OH
|
H — C — OH
|
H

c. CH₃CH(OH)COOH

a. _____   b. _____   c. _____

786

Which of the following is(are) structural isomers? _____

b

a. CH_3CH_2CHO CH_3CH_2COOH

b.
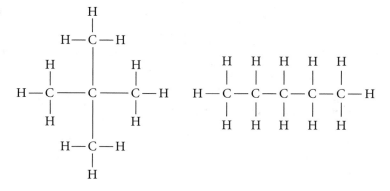

c.
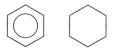

787

a, b

Which of the following is a *cis* isomer? _____ A *trans* isomer?

_____

a.
$$CH_3 - \overset{\overset{\displaystyle CH_3}{|}}{C} = \overset{\overset{\displaystyle CH_3}{|}}{C} - CH_2 - CH_3$$

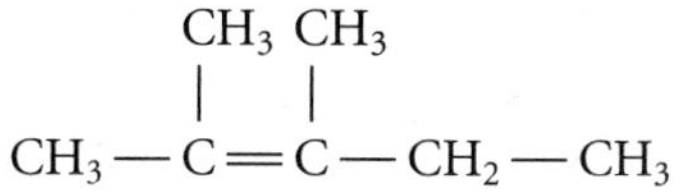

b.
$$CH_3 - \overset{\overset{\displaystyle CH_3}{|}}{C} = \underset{\underset{\displaystyle CH_3}{|}}{C} - CH_2 - CH_3$$

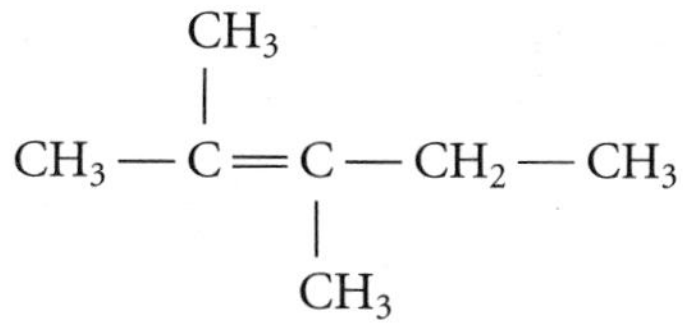

a, b, c

788

Which of the following pairs are enantiomers? _____

a.

```
      O                    O
      ‖                    ‖
      C — H           H — C
      |                    |
HO — C — H          H — C — OH
      |                    |
 H — C — OH        HO — C — H
      |                    |
 H — C — H          H — C — H
      |                    |
      H                    H
```

b.

```
      H                    H
      |                    |
      C = O           O = C
      |                    |
HO — C — H          H — C — OH
      |                    |
 H — C — H          H — C — H
      |                    |
      H                    H
```

c.

```
      O                      O
      ‖                      ‖
      C — H             H — C
      |                      |
 H — C — NH₂     H₂N — C — H
      |                      |
 H — C — H          H — C — H
      |                      |
      H                      H
```

d.

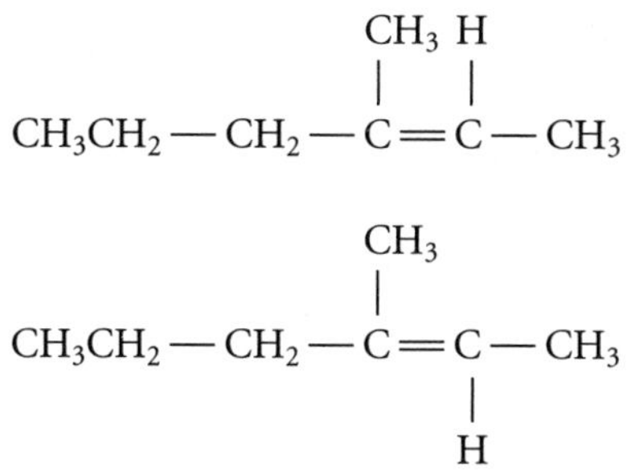

$$CH_3CH_2 — CH_2 — \overset{\overset{\displaystyle CH_3}{|}}{C} = \overset{\overset{\displaystyle H}{|}}{C} — CH_3$$

$$CH_3CH_2 — CH_2 — \overset{\overset{\displaystyle CH_3}{|}}{C} = \underset{\underset{\displaystyle H}{|}}{C} — CH_3$$

789

4

Asymmetric carbons have how many different groups attached to them?

_____

790

3

How many asymmetric carbons are present in the following compound?

_____

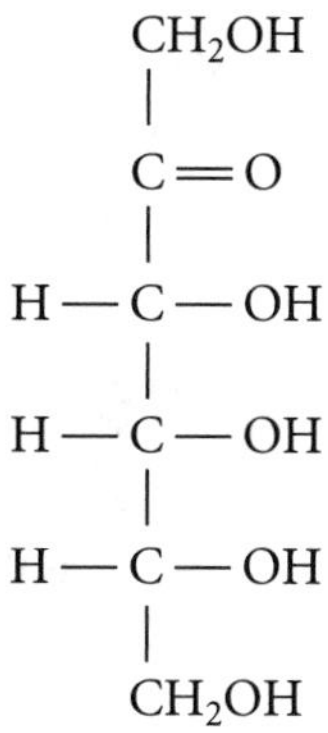

$$CH_2OH$$
$$|$$
$$C = O$$
$$|$$
$$H — C — OH$$
$$|$$
$$H — C — OH$$
$$|$$
$$H — C — OH$$
$$|$$
$$CH_2OH$$

$C_{12}H_{22}O_{11}$

791

791

Which of the following compounds is a carbohydrate?

___ C_6H_6 ___ $C_{12}H_{22}O_{11}$
___ C_6H_7ON ___ C_2H_6O

a. *ose*
b. *ase*

792

a.   Monosaccharides have names ending in _____.

b.   Most enzymes have names ending in _____.

ribose; glucose

793

An example of a pentose is _____. An example of a hexose is

_____.

with one less oxygen

794

The prefix *deoxy* means _____.

two monosaccharides

795

Hydrolysis of a disaccharide yields _____.

796

The bond holding the two halves of a disaccharide together is called a(n) _____ bond.

797

Starch is a _____ (monosaccharide/disaccharide/polysaccharide).

798

a. Which type of carbohydrate can be absorbed directly into the blood? _____

b. Which type of carbohydrate forms a colloidal dispersion in water? _____

799

Which are unsaturated, fats or oils? _____

800

The products of fat hydrolysis are _____ and _____.

glycosidic

polysaccharide

a. monosaccharides
b. polysaccharides

oils

fatty acids; glycerol

fatty acids; glycerol; phosphoric acid; a nitrogen compound

801

A phospholipid, on hydrolysis, yields _____, _____, _____, and _____.

cholesterol or bile salts or sex hormones

802

An example of a steroid is _____.

C, H, O, N

803

Proteins always contain which elements? _____

amino acids

804

The hydrolysis of protein yields _____.

many; peptide

805

A polypeptide contains _____ amino acids held together by _____ bonds.

806

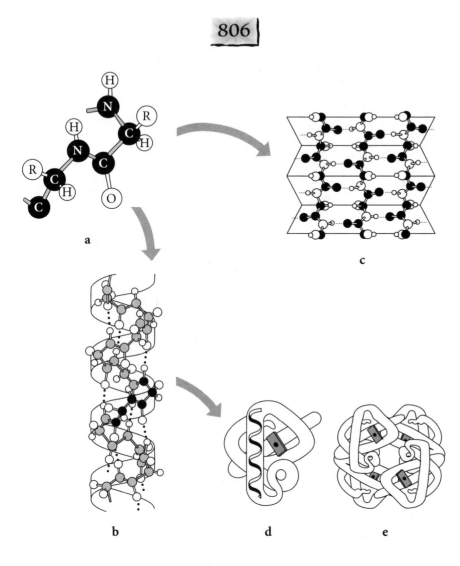

a

c

b   **d**   **e**

Which figure(s) represents the primary structure of a protein?

_____

Which figure(s) represents the secondary structure of a protein?

_____

Which figure(s) represents the tertiary structure of a protein?

_____

Which figure(s) represents the quaternary structure of a protein?

_____

3; peptide

807

A tripeptide consists of _____ (how many?) amino acids linked together by what type of bond? _____

ribose; deoxyribose

808

Nucleotides may contain one of two pentoses. They are _____ and _____.

ATP (or adenosine triphosphate)

809

The compound that provides chemical energy for the body is _____.

2

810

ATP contains how many high-energy phosphate bonds? _____

hydrogen

811

The opposite sides of the DNA coils are held together by _____ bonds.

812

a. In DNA, the partner of A is _____.

b. As RNA is synthesized, the presence of A in DNA will specify the
    incorporation of _____ in the RNA.

c. As RNA is synthesized, the presence of C in DNA will specify the
    incorporation of _____ in the RNA.

d. In DNA, the partner of C is _____.

a. T
b. U
c. G
d. G

813

Complete the following diagram indicating partners and bonds in the
formation of mRNA.

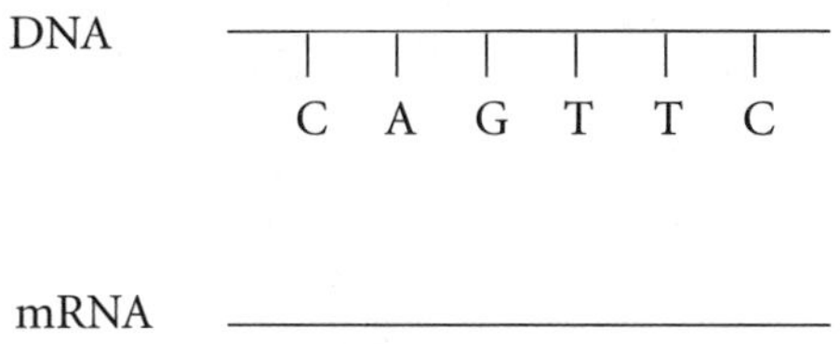

DNA
C   A   G   T   T   C

mRNA   _____

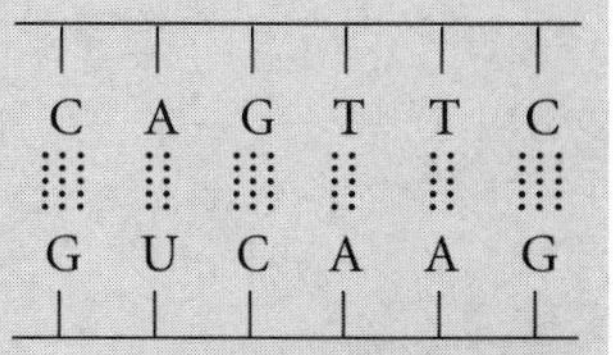

```
C   A   G   T   T   C
⋮⋮⋮ ⋮⋮  ⋮⋮⋮ ⋮⋮  ⋮⋮  ⋮⋮⋮
G   U   C   A   A   G
```

814

Which amino acids are designated by the mRNA in problem 813? (See
table on page 211.) _____

val, lys

mutation

no

specific

coenzyme

a. competitive
   inhibitor
b. noncompetitive
   inhibitor

815

If a genetic code is miscopied, such a change is called a(n) _____.

816

Are all mutations harmful? _____

817

Enzymes are _____ (specific/nonspecific).

818

The nonprotein part of an enzyme is called the _____.

819

a. A substance that "fits" into the active site of an enzyme and prevents
    that enzyme from functioning is called a _____.

b. A substance that alters the shape of the "active" site of an enzyme
    by reacting with that enzyme at a spot other than the active site is
    called a(n) _____.

820

Which of the two acids in the diagram is oxidized? _____

Which substance is reduced? _____

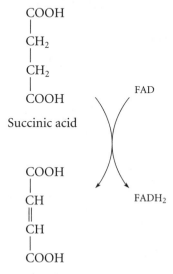

COOH
|
CH$_2$
|
CH$_2$
|
COOH

Succinic acid

FAD

COOH
|
CH
‖
CH
|
COOH

Fumaric acid

FADH$_2$

succinic acid; FAD

821

Glycolysis involves the conversion of glucose to _____.

pyruvic acid

822

Glycolysis takes place in which part of the cell? _____

cytoplasmic fluid

823

Is glycolysis an aerobic or anaerobic sequence? _____

anaerobic

yes

824

Does glycolysis produce ATP? _____

electron transport chain

825

The NADH produced during glycolysis is transferred to which sequence? _____

acetyl CoA;
mitochondrion

826

Pyruvic acid is converted into _____ in which part of the cell? _____

Krebs; mitochondrion

827

Acetyl CoA enters the _____ cycle, which takes place in which part of the cell? _____

anaerobic

828

Is the Krebs cycle an aerobic or anaerobic sequence? _____

yes

829

Is ATP produced during the Krebs cycle? _____

830

The NADH and FADH$_2$ produced during the Krebs cycle are transferred to which sequence? _____

electron transport chain

831

The electron transport chain occurs in which part of the cell? _____

mitochondrion

832

Is the electron transport chain an aerobic or anaerobic sequence?

_____

aerobic

833

In which parts of the cellular respiration sequence is ATP produced?

_____

glycolysis, Krebs cycle, electron transport chain

834

What is the source of energy for photosynthesis? _____

light

835

Which reactants are involved in photosynthesis? _____

carbon dioxide, water

836

From which compound is oxygen produced? _____

water

I; II

837

Which photosystem can be cyclic? _____ Which is noncyclic? _____

ATP, NADPH, O_2

838

Which compounds are produced in the light reactions? _____

ADP

839

ATP is produced from which compound? _____

$NADP^+$

840

NADP is produced from which compound? _____

carbon dioxide

841

Which compound enters the Calvin cycle? _____

glyceraldehyde-3-P

842

Which compound is produced by the Calvin cycle? _____

ADP, $NADP^+$

843

Which high-energy compounds are produced by the Calvin cycle? _____

844

Is oxygen produced in light reactions photosystem I? _____ In light reactions photosystem II? _____

845

Gases flow by diffusion from an area of _____ (higher/lower) partial pressure to an area of _____ (higher/lower) partial pressure.

846

In the lungs, arterial blood picks up _____ gas and gives off _____ gas.

847

What substance in a red blood cell transports oxygen? _____

848

When hemoglobin reacts with oxygen, what compound is formed? _____

849

Is hemoglobin recycled? _____

no; yes

higher; lower

oxygen; carbon dioxide

hemoglobin

oxyhemoglobin

yes

carbonic acid

850

Carbon dioxide is a waste product produced in the tissues. It flows into the red blood cells, where it reacts with water to form _____.

bicarbonate; hydrogen

851

Carbonic acid ionizes to form the _____ ion and the _____ ion.

chloride shift

852

The flow of bicarbonate ions from the red blood cells into the blood plasma and the opposite flow of chloride ions from the plasma into the red blood cells are called the _____.

in the arterial blood in the alveoli

853

Where in the body does the reverse chloride shift take place?

_____

bicarbonate; carbamino-hemoglobin

854

Carbon dioxide is carried in the blood in two forms, the _____ ion and the compound _____.

Conclusion

Congratulations! You have completed the program. If you have worked conscientiously, you should now be able to

- recognize the elements present in various biological compounds

- understand the term *pH* as applied to fluids and cells

- recognize what is meant by the terms *acids*, *bases*, and *salts* as they occur in plant and animal tissues

- know what electrolytes are so that their functions in the life of the cell can be more clearly understood

- recognize organic functional groups

- recognize types of isomers

- understand oxidation and reduction as they occur in the metabolic processes of plants and animals

- differentiate among carbohydrates, lipids, and proteins, and recognize each from its unique functional groups

- understand how enzymes function

- know what nucleic acids are and how they are held together

- understand diffusion and osmosis

- distinguish between passive and active transport

- understand osmotic pressure

- understand transmission of nerve impulses, including depolarization and repolarization

- understand how neurotransmitters work

- know how DNA replicates

- know how mRNA is formed and how it regulates protein synthesis

- understand reactions involved in photosynthesis, including light reactions and the Calvin cycle

- follow the flow of oxygen from the lungs to the cells and the flow of carbon dioxide from the cells to the lungs based on gas partial pressure gradients

Appendix

Figure A
Cellular Respiration

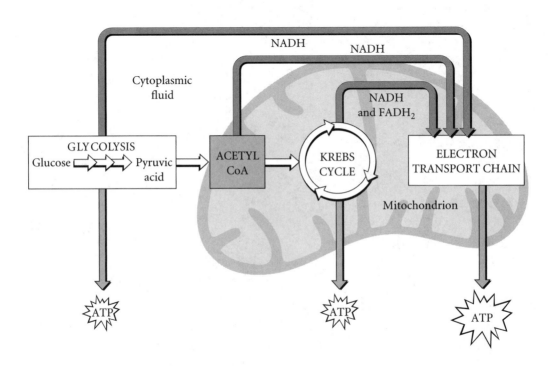

Figure B
Glycolysis

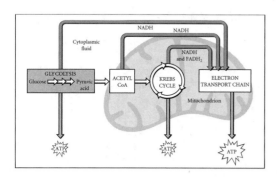

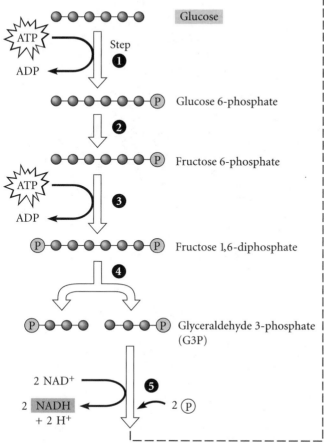

Glucose

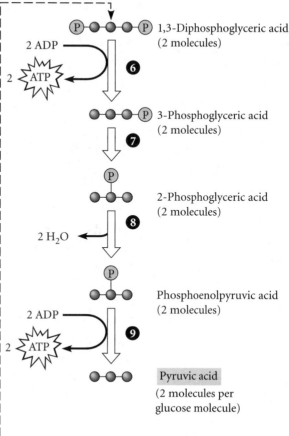

Step **1**

ATP

ADP

Glucose 6-phosphate

2

Fructose 6-phosphate

ATP

ADP

3

Fructose 1,6-diphosphate

4

Glyceraldehyde 3-phosphate
(G3P)

2 NAD⁺

2 NADH
+ 2 H⁺

5

2 Ⓟ

2 ADP

2 ATP

6

1,3-Diphosphoglyceric acid
(2 molecules)

3-Phosphoglyceric acid
(2 molecules)

7

2-Phosphoglyceric acid
(2 molecules)

2 H₂O

8

Phosphoenolpyruvic acid
(2 molecules)

2 ADP

2 ATP

9

Pyruvic acid
(2 molecules per
glucose molecule)

Figure C
Conversion of Pyruvic Acid
to Acetyl Coenzyme A

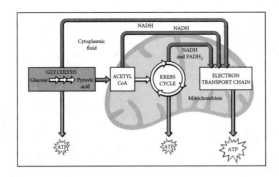

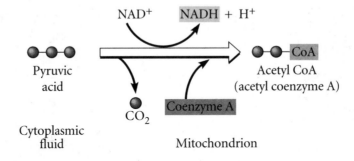

Figure D
The Krebs Cycle

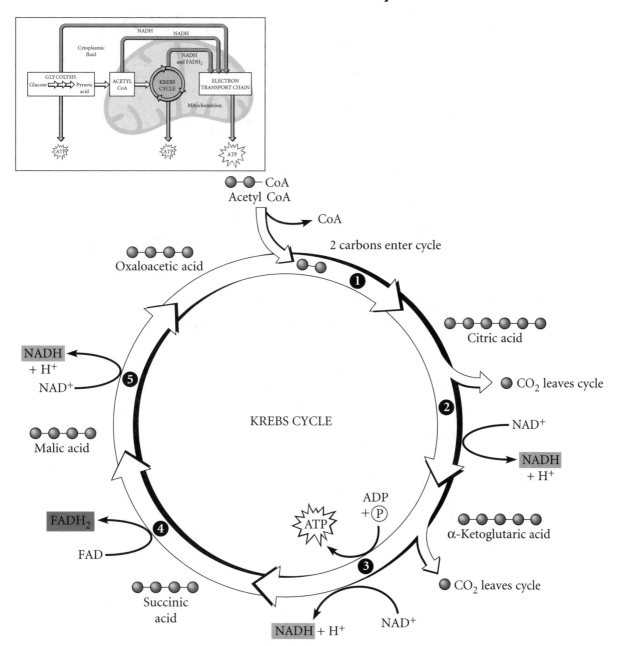

Figure E
Electron Transport Chain

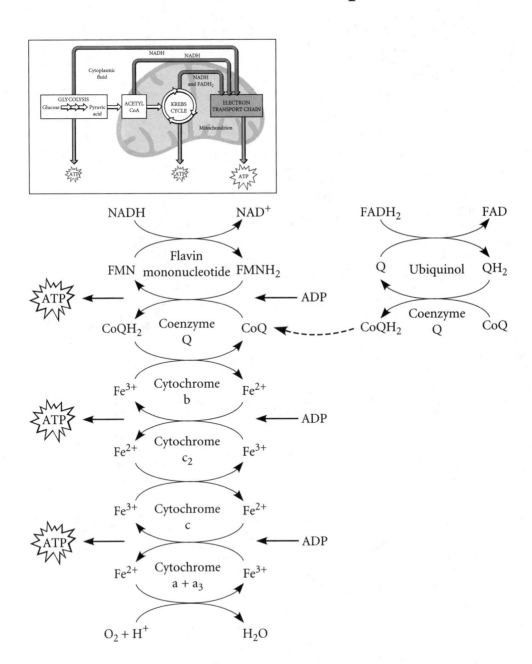

Index